T-LEVELS
THE NEXT LEVEL QUALIFICATION

DIGITAL PRODUCTION, DESIGN & DEVELOPMENT

CORE

Mo Everett
Sonia Stuart

Although every effort has been made to ensure that website addresses are correct at time of going to press, Hodder Education cannot be held responsible for the content of any website mentioned in this book. It is sometimes possible to find a relocated web page by typing in the address of the home page for a website in the URL window of your browser.

Hachette UK's policy is to use papers that are natural, renewable and recyclable products and made from wood grown in well-managed forests and other controlled sources. The logging and manufacturing processes are expected to conform to the environmental regulations of the country of origin.

Orders: please contact Hachette UK Distribution, Hely Hutchinson Centre, Milton Road, Didcot, Oxfordshire, OX11 7HH. Telephone: +44 (0)1235 827827. Email education@hachette.co.uk Lines are open from 9 a.m. to 5 p.m., Monday to Friday. You can also order through our website: www.hoddereducation.co.uk

ISBN: 978 1 3983 4678 9

© Maureen Everett and Sonia Stuart 2023

First published in 2023 by
Hodder Education,
An Hachette UK Company
Carmelite House
50 Victoria Embankment
London EC4Y 0DZ

www.hoddereducation.co.uk

Impression number 10 9 8 7 6 5 4 3 2

Year   2027 2026 2025 2024 2023

Cover photo © Seventyfour – stock.adobe.com

Typeset in India

Produced by DZS Grafik, Printed in Slovenia

A catalogue record for this title is available from the British Library.

Contents

Guide to the book

Learning outcomes

Summaries of the knowledge outcomes that you need to learn in each content area.

Key term

Definitions of key terms.

Industry tip

Tips and advice to help you in the workplace.

Important point

Important points that you need to be aware of.

Activity

Short activities that encourage you to apply the knowledge and skills covered in the Student Book.

Research

Research-based activities – either stretch-and-challenge activities, enabling you to go beyond the course, or industry placement-based activities encouraging you to discover more about your placement.

Case study

Scenarios that place content into real-world contexts.

Test yourself

Short questions designed to test your knowledge and understanding.

Assessment practice

Knowledge-based practice questions to help you to prepare for the core exams.

Skills practice

Short scenarios and focused activities that allow you to apply the skills you have learned in each content area.

Content area 1: Problem solving

Problem-solving skills can be used to analyse problems. This will lead to solutions which can then be developed into code for digital systems. Computational thinking, which you are going to explore in this content area, provides a framework for the ways in which a problem can be solved.

You will also learn about the different approaches that can be applied to problem solving and how these can lead to the creation of algorithms to define the processes. There are different methods of creating algorithms including flowcharts, written descriptions, pseudocode and program code. You will learn how they can be used to create algorithms, as well as exploring the advantages and disadvantages of each method. You will also learn about the different representations of each algorithmic method and how to use them to create an algorithm that accurately represents a problem.

It is very important that algorithms are correct, solve the problem and produce the correct, and expected, output. You will learn how a visual check and trace tables can be used to ensure that the created algorithm is fit for purpose, providing a clear and robust design to inform the coding of the solution to the problem.

Learning outcomes

In this content area you will learn about:
1.1   Computational thinking
1.2   Algorithms

1.1 Computational thinking

By applying **computational thinking** to an overarching problem, you can understand what the problem is, and what it consists of, to help you develop a possible solution. The solutions to the problem can be shown in ways that a digital system, a person, or both, can understand.

1.1.1 Using top-down, bottom-up and modularisation approaches to solve problems

When solving problems related to the use of digital systems there are three main approaches that can be taken:

▶ top-down
▶ bottom-up
▶ modularisation.

Top-down approach

The top-down approach is a technique used to solve problems where the problem is broken down into smaller and smaller problems, modules, until an easily solved problem is defined.

This means that the top-down approach divides a complex problem into multiple smaller parts which can then be used to code the associated modules. Each module is **decomposed** until the final module(s) cannot be further decomposed. This approach uses a stepwise process to break a large problem into simpler and smaller problems, modules, to organise and code the software program in an effective and efficient way. The flow of control in this approach is always in the downward direction.

The top-down approach is usually represented as a tree structure as shown in Figure 1.1.

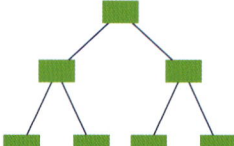 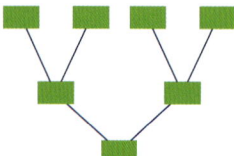

▲ Figure 1.1 A top-down approach (left) and a bottom-up approach (right)

Each level in the top-down approach shows a different level of detail (**abstraction**) with the top level, level 1, showing the greatest level of abstraction. Using the top-down approach, it is possible to break a problem into detailed sub-problems. The top-down approach begins with the abstract problem and refines the problem by decomposition until no more decomposition can be carried out.

> ### Key terms
>
> **Computational thinking:** a problem-solving method using computer science techniques, where possible solutions are developed and presented in a way that can be understood by humans and computers.
>
> **Decomposition:** breaking a complex problem into smaller sub-problems.
>
> **Abstraction:** the process of removing or filtering characteristics that are not needed, in order to focus on essential characteristics.

Bottom-up approach

The bottom-up approach is the opposite of the top-down approach (Figure 1.1). The process starts with the smallest part, module, of the problem. These are then combined to move up a level and this then carries on until the complete problem is solved. The combination of the modules is called **integration**.

Table 1.1 shows the main differences between the top-down and bottom-up approaches.

	Top-down	Bottom-up
Concept	Splitting (Breaks the massive problem into smaller sub-problems)	Merging (Solves the fundamental low-level problems and integrates them into a larger one)
Redundancy	Contains redundant information	Redundancy can be eliminated
Programming languages	Structure/procedural-oriented programming languages (e.g. C)	Object-oriented programming languages (e.g. Python, Java)
Main use	Module documentation, test case creation, code implementation and debugging	Testing

▲ Table 1.1 Features of top-down and bottom-up approaches to problem solving

Modularisation approach

The modularisation approach aims to break a problem into different components, or modules. Most problems are not one big problem but are a collection of different tasks, modules, which are independent and can be separated. A big problem can seem daunting but by breaking it into smaller tasks, modules, the problem can become more manageable and solvable.

All these modules work together to provide a solution to the initial problem. A module can be seen as a subprogram where the main program 'calls' each module. This is a strategy often utilised in object-oriented programming languages.

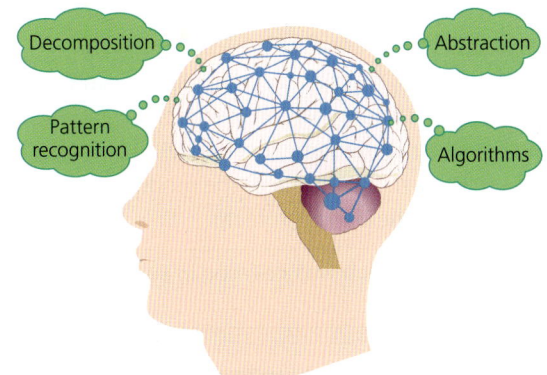

▲ Figure 1.2 The four techniques, or pillars, of computational thinking

> **Activity**
>
> Split into three groups. Each group should choose a different problem-solving approach: top-down, bottom-up or modularisation.
>
> You want to make a cooked dessert as a treat for your family. Using your group's chosen approach, produce a document showing the steps that should be taken.
>
> Each group should then present their solutions.

> **Test yourself**
>
> 1 What is computational thinking?
> 2 How is the top-down approach usually represented?
> 3 What does the top-down approach begin with?
> 4 What type of programming languages use the bottom-up approach?
> 5 What is the combination of modules called?
> 6 What is the aim of the modularisation approach?

1.1.2 Decomposing problems

There are four techniques, also known as pillars, that can be utilised as part of computational thinking. These are:

- ▶ decomposition
- ▶ pattern recognition
- ▶ abstraction
- ▶ algorithms.

Decomposition is one of the techniques involved in the process of computational thinking.

Decomposing a problem is the technique of breaking a complex problem or system into smaller, more manageable separate parts. These can also be called **modules** when the solution is being created.

Once the problem has been broken down, its separate parts can be understood independently. This also means that the parts can then be developed separately to solve the initial problem. Each part can also be evaluated, or tested, when a program has been developed. Using decomposition makes complex problems easier to solve, and large digital systems easier to design and create.

Solving a complex problem as a whole may seem very difficult. However, the solution to each decomposed part may be much simpler. When all the parts are solved, these solutions can be put together. This will provide a solution to the initial problem.

Decomposing problems is a skill that is needed in many different job roles including project management and software design. When a project is being planned, the initial project is broken down into many different subtasks which have to be completed to complete the project.

> **Activity**
>
> You have been asked to plan a trip to Bletchley Park for your teaching group. Split into smaller groups of three or four. Plan the trip and create a digital communication to show all the tasks and subtasks which would be needed. Each group should present their plan.
>
> Was everything included to make the trip a success? Identify any tasks or subtasks that were forgotten, considering the impact these omissions would have on the trip.

The top-down approach to problem solving uses decomposition. This is because the initial problem is broken down until no more decomposition can be carried out. When decomposition is complete, each sub-problem should be at the same level of detail and able to be solved on its own. Once coded, the modules are then combined to solve the initial problem.

> The top-down approach was covered in section 1.1.1 of this content area.

Advantages and disadvantages of decomposition

The advantages of using decomposition to analyse, plan and create a digital solution include that:
- Different people can work on the different sub-tasks, modules, which can then be integrated to make the final solution.
- Maintenance of the final software solution can be completed at modular level.

The disadvantages of using decomposition to analyse, plan and create a digital solution include that:
- The sub-problem modules may not combine to solve the initial problem.
- If the initial problem is not fully understood, then it can be difficult to decompose.

The four steps of decomposition

When decomposing a problem ready for coding there are four steps that should be carried out:
1 **Identify and describe the problems and processes** – the problems should be identified and described. It is important that at this stage the technical terms used should match the sector the problem relates to. For example, if the problem relates to the retail sector then terms related to the retail sector should be used.
2 **Break down the problems into separate tasks** – this stage begins to decompose the problem. It is usual to use the top-down approach as the problem is known. There is no limit on the number of tasks and subtasks that can be identified. What is important is that the problem is decomposed until it can be decomposed no further. It is also important that the final decomposition should only include tasks and subtasks that fully relate to the initial problem.
3 **Describe the tasks and subtasks** – this stage requires documentation to be created that relates to the decomposition. The documentation produced should be clear and concise to enable a third party to implement the solution to the problem.

4 **Communicate** – most software solutions are coded by a team of people. When a problem has been decomposed it is possible that each person will take responsibility for coding a specified task or sub-tasks – also known as modules. When the modules have been coded then they can be bolted together to provide a solution to the given problem.

> You will learn about the different sectors of business in Content area 5, section 5.1.1.

Test yourself

1 What is decomposition?
2 What are modules?
3 How does the top-down approach use decomposition?
4 Identify one advantage and one disadvantage of decomposition.

1.1.3 Using pattern recognition

When a problem has been decomposed into the smallest sub-problems, it is often possible to identify patterns. Patterns can be found everywhere. For example, every school and college has a timetable showing lessons, teachers, rooms, days and times. This is an example of a pattern. This is because the timetable runs for an academic year and does not change over the year.

Research

To understand pattern recognition, it is best to start by thinking about things in a non-digital context. Investigate the artist M.C. Escher. How does the artist use patterns in their work?

The identification and recognition of patterns – things that are common between problems or programs – is one of the pillars of computational thinking. Patterns can help to solve complex problems more efficiently.

There are patterns that are used in many situations. For example, Morse Code was used during World War 2 and for communication between ships.

Research

Binary uses a pattern of 1s and 0s to represent each letter of the alphabet. Find the patterns that are used in binary to represent each letter. Write a phrase in binary. Share your phrase with your group and ask them to decipher it.

Continuing work done by Polish code breakers shared with the UK, the code breakers at Bletchley Park, including Alan Turing, deciphered the coded messages which had been ciphered by the Enigma Machine. The Enigma Machine used different ciphers to convert each message into code. Each cipher had a pattern and it was the recognition of these patterns that provided the initial breakthrough when identifying the cipher used for each message.

Research

Research the Enigma Machine and how the cipher was cracked by Alan Turing and the team at Bletchley Park. Create a digital communication to present the results of your research.

Pattern recognition can also be used in a range of emerging technology applications. For example, **facial recognition**, voice recognition software and automated transport. These applications use predefined and pre-learned patterns to process the inputs and produce outputs.

Key term

Facial recognition software: software that can identify or confirm someone's identity using their face in a photo, video or in real time.

Research

Select an emerging technology application. Investigate how the application uses patterns and its advantages and disadvantages. Discuss your findings with the rest of your teaching group.

By identifying and recognising patterns, it is possible when coding to locate a pre-existing module of code.

There are five main steps to identifying and recognising patterns when beginning to code.

1 **Identifying and interpreting common elements in problems or systems** – when a common pattern has been identified, there is more than likely going to be an existing solution to the problem. For example, a search can be carried out for a specific customer in a customer database. The process of searching a database varies slightly depending on the type of database, the number of records or the purpose of the database.

2 **Identifying and interpreting common differences in problems or systems** – using the customer database example, all customer databases store records of customers. What information is held about the customers and how the information is recorded may be different, but the purpose of the databases is the same – to store information.

3 **Identifying individual elements in the patterns** – the elements can be input, process or output. For example, the customer database may record the number of times a customer has placed an order. These orders may be recorded as a number or by recording the date each order was placed. The inputs are different but the process of calculating the number of times is the same.

4 **Describing patterns that have been identified** – when a pattern has been identified, then it needs to be described. The pattern may be new or one that occurs several times.

5 **Making predictions based on identified patterns** – when a pattern has been identified then a decision can be made about using it multiple times in code or reusing it in a different program.

When the patterns have been identified, it is usually only the specific details that need to be changed within any module of code.

Test yourself

1 What is pattern recognition?
2 What would be the result of not identifying patterns when starting to code?
3 How do emerging technology applications use pattern recognition?
4 Identify and describe one step used in identifying and recognising patterns.

1.1.4 Using abstraction

Identifying necessary information and filtering out the unnecessary

Abstraction is the process of removing or filtering characteristics that are not needed, in order to focus on essential characteristics.

A well-known example of abstraction is the London Underground map.

▲ Figure 1.3 The London Underground map is an iconic visual representation used by millions of people to plan their journeys

By looking at the map, it is possible to plan a journey, know how many stations will be visited and if changes of line need to be made and where. It is not necessary to know how many miles the journey will take or where each station is in relation to the next – these have been filtered out of the map during the process of abstraction.

By carrying out abstraction it enables a general idea of what a problem is and how it can be solved. The process removes specific detail and patterns that do not help you to solve the problem. If abstraction is not carried out, then it is possible that an incorrect solution to the problem may be provided.

Abstraction provides a general idea of the problem rather than focusing on specific details. This general idea is known as a model. By carrying out abstraction the complexity can be reduced while the efficiency can be increased.

Table 1.2 shows the difference between specific and general detail, using the example of making a casserole.

General	Specific
A casserole needs ingredients.	It is not required to know what ingredients.
Each ingredient has a required quantity.	It is not required to know the quantity of each ingredient.
A casserole needs to be cooked for a long time.	It is not required to know the cooking time.

▲ Table 1.2 General versus specific

Creating layers of abstraction

Each layer of abstraction hides the complexity of the layer below. This means that the top abstraction layer hides all the complexity of the problem. There are two main steps involved in abstraction.

1 The information needed to solve a given problem needs to be identified as, without this, the solution may not solve the problem. It is also important to know and understand why this information is needed. The required format of the information should also be considered.

2 Carry out abstraction to filter out the unrequired information. This means that only the information required will be considered. Anything else will be a distraction to the process. It is also important to know and understand why information is required and not required.

Each layer of abstraction needs to be complete. This means that each layer should show:

▶ the inputs – what is inputted into the digital system at that level. This could include validation, and the format of the input, for example currency shown to two decimal places

▶ the outputs – how and in what format the output will be given. For example, a printed document showing specified data and information or on-screen output. This is usually specified by the client

▶ **variables** – these are values that will change. Variables usually change as a result of an input by an end user or of a calculation being carried out

▶ **constants** – these are values that do not change. A constant could be, for example, a fixed delivery price

▶ key processes – these are the actions, processes, that the layer must carry out

▶ repeated processes – these are processes that are carried out several times in a digital system.

Key terms

Variables: in the context of abstraction, these are values that will change usually as a result of an input or of a calculation being carried out.

Constants: values in a program that do not change when the program is being executed or run.

Activity

Create the lowest abstraction level for a digital system that can be used to calculate the costs of a school prom, including inputs, outputs, variables and constants. The output of the system should be the cost of a ticket to the prom and how much profit would be made.

Test yourself

1 Why is abstraction important?
2 What is the aim of abstraction?
3 What happens to the efficiency when abstraction is carried out?
4 Identify one step carried out in the process of abstraction.
5 What is a variable?

1.2 Algorithms

You are going to explore the different approaches that can be applied to problem solving and how these can lead to the creation of algorithms to define the processes. There are different methods that can be taken to create algorithms. You will learn about how flowcharts, written descriptions, pseudocode and program code can be used to create algorithms, including the advantages and disadvantages of each.

Each algorithmic method has defined representation. There are, for example, a set of symbols that are used when creating a flowchart. You will learn about the different representations that are used and how to use these to create an algorithm that accurately represents a solution to a problem.

It is very important that algorithms are correct, solve the problem and produce the correct, and expected, output. You will learn how a visual check and trace tables can be used to ensure that the created algorithm is fit for purpose and, when combined with the other pillars of computational thinking, provides clear and robust design to inform the coding of the solution to the problem.

1.2.1 What algorithms are and how they are expressed

1.2.4 The purpose of a given algorithm and how it works

An algorithm is defined as a plan, or a well-defined set of step-by-step instructions, to solve a problem.

Algorithms are the basis on which all software is created. An algorithm is language independent. This means that step-by-step instructions can be implemented with the expected output being the same irrespective of which language is used.

An algorithm must:
▶ be clear with clearly defined steps
▶ have clearly defined inputs and outputs
▶ be simple, generic and practical
▶ be language independent.

The advantages of algorithms include:
▶ They are easy to understand by anyone.
▶ They are a step-by-step representation of a solution to a given problem.
▶ The initial problem is broken down into steps, which means it is easier to convert into code.

The disadvantages of algorithms include:
▶ Creating a complete algorithm can be time consuming.
▶ Some constructs can be difficult to represent.

An algorithm can be represented as a flowchart, written description, pseudocode or program code. You are going to explore each of these.

Flowcharts

A flowchart is a graphical diagram that represents an algorithm which can be used to solve a problem.

The advantages of flowcharts include:
▶ The flow of the program can be seen clearly.
▶ Flowcharts are created using a standardised set of symbols so can be interpreted and understood by many people.

The disadvantages of flowcharts include:
▶ With a large, complicated program the flowchart can become very large and difficult to follow.
▶ Changes to the design may result in the flowchart being amended or redrawn.

Written descriptions

Written descriptions are generally written in a natural language or plain English language. Unlike flowcharts, pseudocode and program code, there is no defined format for this method of algorithmic representation. A written description should be easy to understand and contain little, if any, specific detail. It is often better to use very short sentences. It is important, however, that all parts of the final decomposition of the problem are included.

The advantages of using a written description include:
- There is no formal **syntax**, so many people can create a written description.
- It is automatic and natural to use 'proper' English.

The disadvantages of using a written description include:
- the temptation to create a complete description
- failure to include every step required.

> ### Key term
>
> **Syntax:** in the context of algorithms, these are the structure of statements.

Pseudocode

Pseudocode is an informal programming description showing the flow through the process. Pseudocode is written in a format that is similar to the structure of a high-level programming language. Pseudocode provides an outline of what the resulting program should achieve.

Pseudocode has its own syntax, some of which is very similar to many actual programming languages. Pseudocode algorithms will not run unless they are converted into an actual programming language.

The advantages of using pseudocode include:
- The pseudocode can be converted into programming code with minor changes to obey the syntax rules of the programming language.
- It can be easy to follow and understand even if errors are present in the pseudocode.
- Unlike using a flowchart, changes can be implemented quickly.
- It can act as a link between the algorithm and the final program.
- It explains the purpose of each line of code, so if the pseudocode is fully complete and detailed, the creation of the final code should be uneventful and meet the needs of the client.

The disadvantages of using pseudocode include:
- It can be as time consuming to write clear and well-structured pseudocode as it is to write the final programming code.
- It can be difficult to see the logical flow of the program.

> ### Industry tip
>
> There are many different variants of pseudocode. Each organisation may use a different variant. The pseudocode you should use during this course is defined in Appendix 2 of the qualification specification.

Program code

Program code is usually considered to be the code written to solve the problem which follows the rules of the selected programming language. However, program code can also be used to create an algorithm. In this situation the code is known as 'draft program code'.

Errors can be evident in the draft code as this will not be used to run the final programming code. What is important is that, as with other algorithmic techniques, the meaning is clear and it can be followed.

The advantages of using programming code include:
- It is very probable that the person creating the algorithm will have some programming knowledge.
- It is not necessary to use the correct syntax.
- All the required constructs will be available to be included in the draft code.

The disadvantages of using programming code include:
- It is easy to begin creating draft code but end up creating the final program code.
- Full decomposition may not be completed.

> ### Research
>
> Find one more advantage and disadvantage for each of the algorithmic representations. Discuss your findings with the rest of your teaching group.

Each algorithm will have a purpose which will be the problem to be solved. There are a number of methods that can be used to ascertain the purpose of an algorithm. If the algorithm is simple, then it can be straightforward to determine the purpose. But if the algorithm is more complicated, then a **trace table** can be used. A trace table is a tool that can be used to dry run the algorithm.

Dry running an algorithm means to use values for the variables used in an algorithm and to trace, or run, the processing before beginning to code.

A trace table can allow the values assigned to the variables to be recorded.

> ### Key term
>
> **Trace table:** a tool used to test or dry run algorithms to make sure no logical errors occur while calculations are being processed. Each column represents a variable and the rows represent the numerical input and the output of the variable.

The simple pseudocode algorithm shows the multiplication of a user input up to its 7th value. NUM represents the user input to be multiplied up to the 7th value.

The user inputs a value of 8 for NUM. The trace table shows the values that would be output by the algorithm.

```
RECEIVE num FROM (INTEGER)
KEYBOARD

FOR number FROM 1 TO 7 DO

SEND num * number TO DISPLAY END FOR

OUTPUT

num * number

END FOR
```

NUM	number	Output
8	1	8
	2	16
	3	24
	4	32
	5	40
	6	48
	7	56

By using a trace table, it can be confirmed that the logic and processing in the algorithm is correct.

Many software programs have elements which perform the same task. This means that some algorithms follow patterns that can be recognised. These are usually referred to as standard algorithms. These types of algorithms follow a set pattern and are usually connected with **searching** and **sorting** data.

Key terms

Search: examine data to find a specified value.

Sort: put a data set into a specified order.

Standard searching algorithms include linear and binary searches, while standard sorting algorithms include bubble and merge sorts.

> Searching and sorting methods are covered in Content area 2, section 2.4.6.

Many programming languages have libraries of pre-written and defined code that can be used. This means that algorithms can use standard elements from the libraries.

There are advantages of using code from a library. One is that it saves development time when creating the code. This will also save testing time as the pre-written code will have been used and tested by many different people. Another advantage is that the pre-written code will be optimised, meaning that the code will be efficient and free of any errors.

The main disadvantage of using code from a library is that it may not be exactly what is required. This will mean that the code will need to be edited, so taking more time. It may be quicker, in this case, to write the code from scratch. By doing this, it can be guaranteed that the code will meet the exact requirements.

It is sometimes difficult to integrate pre-written code into other code. You may need to edit to ensure that, for example, any variable names or the data types used are consistent.

> More on using pre-written code in Content area 2, section 2.5.1.

Research

Find standard pseudocode algorithms for a bubble sort and a linear search. Compare your results with the algorithms found by the rest of your teaching group. What are the similarities and differences?

A visual check can also be used to determine the purpose of the algorithm. Looking at the simple pseudocode algorithm it is clear that this solves a maths problem, and that multiplication is involved. This is shown by the use of the * operator and the loop that shows how many numbers should be output before the code stops.

Data types are covered in Content area 6, section 6.2.1.

Activity

Using the pseudocode for the bubble sort and linear search you found, carry out a visual check. For each pseudocode create a written description algorithm. Could the pseudocode be made more efficient?

Test yourself

1 What is an algorithm?
2 Identify one advantage and one disadvantage of using an algorithm.
3 Identify two ways algorithms can be represented.
4 What is meant by the term 'language independent'?
5 What is meant by the phrase 'completing a dry run'?
6 How can a trace table be used to check the logic and processing of an algorithm?

1.2.2 Express an algorithm using flowcharts and pseudocode, and understand how to use these when planning a digital solution

1.2.5 Determine the correct output of an algorithm

1.2.6 Identify and correct errors in an algorithm

A flowchart makes use of a standardised set of symbols which are connected by lines showing the flow of the algorithm. Flowcharts are used to visually show the steps in a process. They should be simple and free from any difficult to understand terminology or technical jargon.

There are many sets of flowchart symbols but the set to be used in this course are shown in Table 1.3.

Pseudocode can use many different sets of syntax. In this course the syntax you will use is provided in the qualification specification in Appendix 2.

The syntax provided shows, for example, data types, constants, data structures, identifiers and comments. Comments can be used to explain the process shown in the pseudocode.

Symbol	Meaning
	The start and end of the flowchart
	A process that has to be carried out
	A sub-process
	A decision – this must have 2 outputs: true/false, yes/no
	An input or output
	A connection – used to link parts of a flowchart that cannot be easily connected, for example when the flowchart goes onto a different page
	The flow of the algorithm – arrows are used to show the direction of the flow

▲ Table 1.3 Symbols used in flowcharts and their meanings

Activity

Create a flowchart to add six numbers together and output the average.

Pseudocode includes a range of arithmetic expressions and relational and logical operators.

Arithmetic expressions are used in calculations. There are four standard expressions:

▶ add
▶ divide
▶ multiply
▶ subtract.

However, the group also includes:

▶ exponent
▶ modulo
▶ integer division.

	Definition	Example expression
Exponent	A quantity representing the power to which a given number or expression is to be raised. This is represented by a ^	$3\wedge 3$ $3*3*3 = 27$
Modulo/ Modulus	The remainder after dividing one number by another. This is represented by mod.	$50\ \%\ 8 = 6$ remainder 2 15 MOD 8 equals 2
Integer division	Division in which the remainder is discarded. This is represented by a /	$50\ /\ 8 = 6$ remainder 2 50 DIV 8 = 6

▲ Table 1.4 More pseudocode expressions and their meanings

Relational operators are also used in pseudocode. They define a relationship between two values. There are six relational operators that are used in this course. These operators can be used for numerical or text-based data.

	Description	Example expression
=	Equal to or equivalent	$x = 9$
<>	Does not equal	$x <> 9$
>	Greater than	$x > 9$
>=	Greater than OR equals	$x >= 9$
<	Less than	$x < 9$
<=	Less than OR equals	$x <= 9$

▲ Table 1.5 The six relational operators that you will use for this course

Logical operators are words used to connect two or more conditions. The output produced depends on the original conditions and on the meaning of the operator. It will usually be 'true' or 'false'. There are three logical operators that are used in this course.

	Description	Example expression
AND	The output will be true if both conditions are true	IF $x > 1$ AND $x < 9$
OR	The output will be true if any of the conditions are true	IF animal = 'dog' OR animal = 'cat'
NOT	The output of the condition will be reversed	WHILE NOT x

▲ Table 1.6 The three logical operators that you will use in this course

It is possible that more than one algorithm will be created for any given program or module.

Over the life of a program or module it is possible that changes will need to be made. It is, therefore, very important that any algorithms are well documented and can be referred to if needed by the person making the changes.

By visual checking and dry running it should be possible to identify any errors in an algorithm at the design stage.

Algorithms are used in many different situations. For example, the price of a book can increase if the sales are high, while the price can decease when sales slow down. The process of moving the price is based on an algorithm.

When applying for a credit card there is a strong probability that an algorithm will decide if the application is successful or not, what the credit limit will be or what the interest rate will be when the credit card is used. By answering application questions, decisions are made that meet the rules of the algorithm. For example, if an application is from someone who has very little credit history, then it is likely that the algorithm will approve the application, subject to status, but with a low credit limit.

Algorithms are also used to provide customers with personalised recommendations based on previous interactions with a business. For example, on a streaming service, recommendations are provided based on previous series or films that have been watched. Large retail websites will provide examples of other items that were purchased by others at the same time as the item being purchased. Recommendations can also be made based on a customer's previous purchases.

Algorithms are also used to make decisions that will affect society. For example, when determining where to put a new hospital or emergency service hub.

1.2.3 Write algorithms that make use of programming constructs

Algorithms make use of sequence, selection and iteration. In a flowchart these can be represented by a decision box and iteration while in pseudocode these can be represented by expressions including:

```
REPEAT
    <command>
UNTIL <expression>
```
And
```
IF <expression> THEN
    <command>
END IF
```

> Before writing algorithms using programming constructs you will need to understand sequence, selection and iteration. You should read Content area 2, section 2.4.1 first.

Skills practice

An online retailer applies delivery costs to each order. The delivery cost is calculated on the total cost of the items bought. The minimum order is £10. The table shows how the delivery costs are calculated.

Cost of items	Delivery cost
£10.00 – £24.99	£2.50
£25.00 – £39.99	£5.00
£40.00 – £59.99	£7.50
£60.00 – £74.99	£10.00
£75.00 or above	Free

The retailer stores the customer delivery addresses, the items they have purchased, the total cost of the items and delivery costs. The retailer wants to implement a digital solution for storing the information and calculating the delivery costs.

You have been asked to:

▶ Create a top-down diagram to show the decomposition of the problem (i.e. the processes carried out by the online retailer).
▶ Create a flowchart to show how the delivery costs will be calculated.
▶ Create pseudocode, based on the flowchart, including the use of comments.
▶ Use a trace table to dry run the pseudocode, making any edits as required. Any remedial edits should be documented.
▶ Explain how the retailer could use pattern recognition.
▶ Explain how algorithms could be used to increase sales.

Assessment practice

1 Explain the difference between the top-down and bottom-up approaches to solving problems.

2 Explain why using decomposition can simplify the solving of a complex problem.

3 Identify and describe two steps to identifying and recognising patterns when beginning to code.

4 Explain the differences between general and specific detail when using abstraction.

5 Compare the use of pseudocode and written description when creating an algorithm.

6 Explain one advantage and one disadvantage of using code from a library.

7 How can trace tables be used to check for errors in an algorithm?

8 Discuss the advantages of using a flowchart to represent an algorithm.

9 Describe how relational operators can be used in algorithms.

10 Explain why algorithms are used in the financial sector.

Content area 2: Introduction to programming

```
var highlight = function($element, pattern)
    if (typeof pattern === 'string' &&
        var regex = (typeof pattern === 'string')

    var highlight = function(node) {
        var skip = 0;
        if (node.nodeType === 3) {
            node.data.search(regex);
            && node.data.length >
            = node.data.match(rege
            ode = document.createEl
            .className = 'highligh
            lebit = node.splitText
            bit = middlebit.splitTe
            ddleclone = middlebit.c
            appendChild(middlecl
            Node.replace
    skip = 1;
```

Programming is the writing of computer code to create a program that will solve a problem. Programs are created to execute algorithms which can be represented as a flowchart or pseudocode. Programming is the translations of these into a computer program which can be understood and acted upon by a computer. Programs must be written in a way that they tell a computer exactly what to do and how to do it.

To write programs, you need to use a programming language. This is an artificial language that the computer understands. It consists of a series of statements which are combined to form instructions that tell the computer what to do.

There are numerous programming languages; some are complex and others are not. In addition, different programming languages work in different ways. For example, in BASIC, the instructions are usually written in uppercase; in Python, they are written in lowercase. While programming languages are designed to be easy for humans to comprehend, a computer is unable to run the programs written in programming language but has to have it translated into machine code.

Learning outcomes

In this content area you will learn about:

2.1	Program data		2.5	Built-in functions
2.2	Operators		2.6	Validation and error handling
2.3	File handling		2.7	Maintainable code
2.4	Program structure		2.8	Testing

2.1 Program data

2.1.1 The use of, and need for, data types

Data can take many forms. There are five main data types that are used in programming. These are:

- Boolean – only two choices, for example true/false, yes/no
- character – a single character which can be a letter, number or symbol
- integer – whole numbers, positive or negative
- real/float – any number, with or without decimal places, positive or negative
- string – a group of characters stored together.

The data type that is used will depend on what processing is to be carried out, the type of data needed by the end user and how the data is stored in a file.

> The Python constructs and explanations of the data types is covered in Table 6.2 in Content area 6, section 6.2.1.

The data type is set when a value is assigned to a constant or a variable. It should not be changed. Any data type that represents a number in Python is known as an **immutable** data type.

> ### Key term
>
> **Immutable:** when the value of a number data type changes, a new object is created.

How the expressions will be constructed will depend on the data type being assigned to a constant or a variable. Table 2.1 shows examples of the expressions for each data type.

Data type	Example expression	Output
Boolean	y = False print(y)	False
Integer	y = 9 print(y)	9
Real/float	y = 45.6 print(y)	45.6
String	y = "Good Morning" print(y)	Good Morning

▲ Table 2.1 Examples of expressions for Boolean, integer, real/float and string data

Character does not have Python notation as this would be a string with a length of 1.

It is possible to check the data type that has been assigned. To do this the code would be:

```
y = 2.3
print(type(y))
```

This expression would return the value <class float>.

> ### Activity
>
> Create a table to show the different data types that would be used in a library (of books). For each data type give an example, explaining why the data type has been used. Discuss your findings with the rest of your teaching group.

> ### Test yourself
>
> 1 Identify two values that can be stored as the Boolean data type.
> 2 What is meant by the integer data type?
> 3 Define the term 'constant'.
> 4 Why does the data type character not have Python notation?
> 5 What is the Python code to check a data type?

2.1.2 Declaring and using constants and variables that use appropriate data types

Python has no command for declaring a constant. (For PEP8 however, the name for a constant is written in uppercase with words being separated by an underscore. This ensures that they are viewed differently from variables, e.g. GROSS_TOTAL.) The variable is created as soon as a value is assigned to it.

For example:

```
j = 9

k = "Monty"
```

This code has assigned the value 9 to the variable j, and Monty to the variable k.

Variables do not need to be set to any specific data type and can, if needed, change type. As defined in section 2.1.1, it is possible to check the data type that has been assigned to a variable using the type() function.

```
j = 9

j = "Misty"

print(j)

print(type(j))
```

The output from this code would be:

```
Misty

<class 'str'>
```

If a specific data type needs to be assigned to a variable, then the code needs to define this. Example code, the data type and the expected output is shown in Table 2.2.

Example code	Data type	Expected output
y = bool(3)	bool	True
y = str("Good Morning")	str	Good Morning
y = int(15)	int	15
y = float(12.5)	float	12.5

▲ Table 2.2 Example code, data type and expected output

There is also a process called casting. When this process is being used the data type functions perform different processes:
- ▶ int() – returns an integer number from an integer, a float (by removing all decimals) or a string (when the string is a whole number)
- ▶ float() – returns a float number from an integer, a float or a string (where the string is a float or an integer)
- ▶ str() – returns a string from a variety of data types.

For example:

```
m = int(3)

n = int(5.7)

p = int("9")
```

The output if these were printed would be:

```
3

5

9
```

Variables can be named. This will make it easier for the code to be understood by a third party and will also help to maintain the code.

> Naming conventions and PEP8 are covered in section 2.7.1 of this content area.

Variables can be either local or global. Whether variables are local or global will depend on their purpose in the program code.

> Local and global variables are covered in section 2.1.4 of this content area.

The choice of data type is very important. The data types that are used in constants and variables will depend on the purpose of the program, the data types of the original data and the requirements of the end users.

For example, integer should not be used for telephone numbers as this data type will remove the leading 0. A telephone number should be stored as a string data type.

2.1.3 The use of, and need for, data structures

Data structures are a collection of data objects which facilitate the storing and managing of data in the program so that it can be used. There are many types of data structures including lists, arrays and dictionaries. Data structures are used for simple and complex computations and are used in all aspects of computer science. They have an important role within algorithms as they enable programmers to manage the data in a more efficient way.

Computer algorithms have increasingly become more complex and the amount of data usage has also increased. The combination of these two things can have an impact on the performance of a program. High processing speed is required to handle vast amounts of data, and searching for data can be challenging. In addition, there may be multiple requests from multiple users happening simultaneously that would be expected to happen instantly (by the end user).

Advantages of data structures

▶ A data structure will optimise memory usage within a system, for example lists and arrays can be linked when the size of the data is uncertain. When there is no further memory available, the data can be released.
▶ Data structures can be reused and compiled into libraries.

Data structures are the foundation of any programming language.

▶ **List** – a list is a collection of data items and can be of different data types, for example numeric, character, logic and so on. If the list is not amended in any way and it is run continuously (e.g. looped), then the outcomes will also be consistent, i.e. they will not change. Lists are changeable (the items within the list can be edited, removed and new items added), and they can contain duplicate values. Each data item within a list is given an **index** value. Lists are one of the four built-in data types in Python.
▶ **Array** – this is a data structure that contains groups of elements. The elements are all the same data type, for example string, character, integer. They are commonly used to organise data so that a related set of values can easily be searched and/or sorted.

▶ **Dictionary** – this is general purpose data structure for storing a group of objects. It is associated with a set of keys and each key has a single associated value. When the dictionary is associated with a key, it will return the associated value linked to the key. Dictionaries support many operations, for example retrieving a value, updating or inserting a value, removing or deleting a key/value combination, and testing or verifying the existence of a key. Items in a dictionary are unordered, so any loops over dictionaries will return the items in a random order.

> Further information on lists can be found in section 2.4.2 of this content area.

Test yourself

1 Define the term 'data structure'.
2 Identify two different types of data structure.
3 Explain one advantage of using a data structure.
4 A dictionary is ordered. True or false?
5 Explain what arrays are commonly used for.

2.1.4 How to manage variables within a program

The scope of a variable relates to where a variable or constant can be accessed within a program. The scope of the variable can either be local to a particular **subroutine** within the program or it can be global to the entire program.

Key terms

Index: a numerical representation of an item's position in a sequence.

Subroutine: sometimes referred to as a routine, a function, a procedure or a subprogram, it is code that is called and executed anywhere in a program. An example is to display a message: instead of writing out the code each time, routines are created and called when required.

Local variable

A local variable is only accessible within a pre-determined part of a program. It is usually declared within a subroutine or is used as an argument that has been passed a value. Local variables that are in different subroutines can have the same identifier (name). The variables are still distinct from each other because the data belonging to each variable is stored in a separate area in **random access memory (RAM)**. Therefore, if you change the value of a local variable within one subroutine, it will not change the value of any other local variable with the same identifier in any of the other subroutines.

Example

```
def my_variable( ):

    x = "local"

    print( x )

my_variable( )
```

Output

```
local
```

> **Key term**
>
> *Random access memory (RAM):* short-term data storage.

Global variable

A global variable takes up more resources than a local variable and must remain in the memory (RAM) the entire time the program is running. This is because it has to be available for access anywhere within a program, at all times. Errors in coding can result in accidental or unintentional changes taking place which can have an impact in another part of the program. Global variables can only be used after they have been declared.

Example

```
x = "global"

def my_variable( ):
    print( "x  inside:" , x)

my_variable( )

print( "x  outside:" , x)
```

Output

```
x inside: global

x outside: global
```

Scope of variables

▶ The scope of a local variable is the subprogram where it has been declared.
▶ The scope of a global variable is the complete program.

Using local and global variables

▶ Global variables must be declared at the start of the program so that they can be used within a subroutine/procedure within the program.
▶ Local variables are declared within subroutines or programming blocks and can only be used within the scope of where they have been declared.

Variable naming conventions

Naming variables is an important aspect to making code readable. A good programmer will create variable names that describe their function and will follow a consistent theme throughout the coding.

Multiword delimited

The convention is to separate words within the name of the variable using 'white space'. This is difficult for many programming languages to interpret. Therefore, the names of variables must be delimited in other ways which are accessible to programming languages. Here are some examples:

▶ **Snakecase** – this is where words are delimited through the use of underscores.
Example:
Variable_a
Variable_b
Variable_c

▶ **Camelcase** – the words in the name of the variable are delimited using capital letters. But it should be noted that the first letter of the variable name does not have a capital letter.
Example:
variableA
variableB
variableC

▶ **Pascalcase** – words are delimited by capital letters (including the first letter of the name).
Example:
VariableA
VariableB
VariableC

▶ **Hungarian notation** – this notation describes the purpose and/or type of variable at the start of the name. It is then followed by a descriptor that states the function of the variable. The Camelcase notation is then used to delimit the words within the name.
Examples:
arrEmployeePayslips // array called "Employee Payslips"
iStudentMarks // integer called "Student Marks"

While a programmer is not required to follow these conventions, it **is** good practice however. When naming variables it is important they are consistent (with respect to format) and easily readable and understandable. It can also depend on the programming language being used, for example a constant variable's name in C++ is all capital letters delimited by an underscore, for example CONSTANT_VARIABLE.

Test yourself

1. Describe the difference between a local and a global variable.
2. What is the scope of a global variable?
3. Write the variable name "This is my name" using the Camelcase notation.
4. Identify the notation that uses underscores for variable naming conventions.
5. Write the code for defining and calling a local variable for a = 7.

2.2 Operators

2.2.1 Mathematical operators in program code and algorithms

Mathematical operators enable calculations to be carried out on any values used in program code.

Mathematical operators, and how they can be used in algorithms, were covered in Content area 1, section 1.2.

The mathematical operators have specific notation in Python code.

Operator	Python notation	Explanation	Example expression
add	+	Adds the values each side of the operator	ans = x + y
subtract	–	Subtracts the value on the right of the operator from that on the left	ans = x – y
multiply	*	Multiplies values on either side of the operator	ans = x * y
divide	/	Divides value on the left of the operator by that on the right	ans = x / y
integer division (DIV)	//	Division in which the remainder is discarded	ans = x // y
Modulus (MOD)	%	Divides value on the left of the operator by that on the right and returns the remainder	ans = x % y

▲ Table 2.3 Mathematical operators and their specific notation in Python code

Using the example notation in Table 2.3, Table 2.4 shows the results of each mathematical operator.

The values are: x = 31, y = 15

Operator	Example	Results
add	x + y	46
subtract	x – y	16
multiply	x * y	465
divide	x / y	2.06
integer division (DIV)	x // y	2
Modulus (MOD)	x % y	1

▲ Table 2.4 The results of each mathematical operator

Taking this forward, the start of the Python code to carry out the examples given in Table 2.4 and print the result would be:

```
x = 31

y = 10

z = 0

z = x + y
```

```
print("Line 1 - value of
     z is  " , z)

z = x - y

print("Line 2 - value of z
     is  " , z)
```

When the above Python code is executed, or run, the output will be:

```
Line 1 — value of z is 46

Line 2 — value of z is 16
```

2.2.2 Relational operators

Relational operators define the relationship between two values.

Relational operators were covered in Content area 1, section 1.2.

Using the example notation in Table 2.5, Table 2.6 shows the results of each relational operator.

The values are: x = 35, y = 25

Operator	Python notation	Explanation	Example expression
Equal to	==	If the two values are the same, the condition is true	x == y
Not equal to	!=	If the two values are not the same, the condition is true	x != y
Greater than	>	If the value on the left is greater than the value on the right, the condition is true	x > y
Greater than OR equals	>=	If the value on the left is greater than, or equal to, the value on the right, the condition is true	x >= y
Less than	<	If the value on the left is less than the value on the right, the condition is true	x < y
Less than OR equals	<=	If the value on the left is less than, or equal to, the value on the right, the condition is true	x <= y

▲ Table 2.5 Example notation

Operator	Example	Results
Equal to	x == y	False
Not equal to	x != y	True
Greater than	x > y	True
Greater than OR equals	x >= y	True
Less than	x < y	False
Less than OR equals	x <= y	False

▲ Table 2.6 The results of each relational operator

Taking this forward, the start of the Python code to carry out the examples given in Table 2.6 and to print the result would be:

```
x = 35

y = 25

if x == y :

     print("Line 1 - x is equal
     to y")

else :

     print("Line 1 - x is not
     equal to y")

if X != y :

     print("Line 2 - x is not
     equal to y")

else :

     print("Line 1 - x is equal
     to y")
```

When this Python code is executed, or run, the output will be:

```
Line 1 - x is not equal to y

Line 2 - x is not equal to y
```

Activity

Create Python code to carry out each of the relational operators. Use values of your choice with the result of each operation being printed on a separate line. Execute your code, correcting any errors.

2.2.3 Boolean operators (NOT, AND, OR)

Boolean operators can also be known as logical operators. Boolean, or logical, operators are words used to connect two or more conditions. The output produced depends on the original conditions and on the meaning of the operator.

Boolean, or logical, operators, and how they can be used in algorithms, were covered in section 2.1.1.

Operator	Python notation	Explanation	Example expression
AND	and	The output will be true if both conditions are true	x > 15 and x < 30
NOT	not	The output of the condition will be reversed	not(x>15 and x<20)
OR	or	The output will be true if any of the conditions are true	x<25 or x>12

▲ Table 2.7 Example notation

Using the example notation in Table 2.7, Table 2.8 shows the results of each Boolean operator.

Taking this forward the start of the Python code to carry out the examples given in the table and print the result would be:

```
x = 25
print(x > 15 and x < 30)
print(not (x>15 and x<20))
print(x > 35 or x < 12)
```

When the above Python code is executed, or run, the output will be:

```
True

True

False
```

Activity

Create Python code to carry out each of the Boolean operators. Use values of your choice with the result of each operation being printed on a separate line. Execute your code, correcting any errors.

Test yourself

1 What is the Python notation for integer division (DIV)?
2 What is the purpose of the Modulus (MOD) mathematical operator?
3 What is the notation for the greater than OR equals operator?
4 What are Boolean operators also known as?
5 What is the purpose of the not Boolean expression?

Operator	Example	Results	Explanation
and	x > 15 and x < 30	True	Both of the conditions are true
not	not (x>15 and x<20)	False	Both conditions are true, but False is returned as the not operator reverses the outcome
or	x > 35 or x < 12	False	Neither of the conditions are true

▲ Table 2.8 The results of each Boolean operator

2.3 File handling

2.3.1 Text files for input and output of data

Every programming language supports file handling. This means that when a program is being executed or run, files can be:

▶ created
▶ opened
▶ read
▶ written to/edited/appended
▶ deleted.

Python has functions that can be used when file handling is being carried out. Python treats files as text or binary (e.g. images) depending on the contents of the file. Python defaults to text files. The Python notation for declaring the type of file is given in Table 2.9.

File type	Python notation	Explanation
Text	"t"	Python's default file type
		Text mode
Binary	"b"	Binary mode
		Example – images

▲ Table 2.9 Python notation for declaring the type of file

> Built-in functions are covered in section 2.5 of this content area.

To open a file in Python the code is:

```
open()
```

The function has two **parameters** which must be declared. These are the file name and mode.

Key terms

Parameter: a special kind of variable in computer programming languages that is used to pass information between functions or procedures.

Append: adding additional data to an existing file.

There are four different modes (methods) of opening files in Python. These are given in Table 2.10.

Mode (method)	Python notation (parameter)	Explanation
Append	"a"	Opens a file for editing – **appending**
		The file is created if it does not exist
Create	"x"	Creates the specified file
		An error is returned if the file already exists
Read	"r"	Python's default value
		A file is opened for reading
		An error is returned if the file does not exist
Write	"w"	Opens a file for writing
		The file is created if it does not already exist

▲ Table 2.10 The four different modes (methods) of opening files in Python

Opening and reading a file

To open and read a file the name of the file should be given. In the example Python code the file name has been given:

```
h = open("practicefile.txt", "rt")
```

As Python defaults to reading a file and a text type file there is no need to define, in the code, that the file is to be read, "r" or the file type, "t". This means that the code shown will be acceptable to Python to open and read the text file 'practicefile'.

```
h = open("practicefile.txt")
```

If the file to be read is stored in the Python folder then the code to read and print the file would be:

```
h = open("practicefile.txt", "r")

print(h.read())
```

The inclusion of the "r" in the code is optional as Python defaults to read.

If the file is stored in a different location, then the code to read and print the file would be:

```
h = open("D:\\myfiles\practicefile.txt", "r")

print(h.read())
```

Activity

Create and save a text file using the text:

> This is my practice file.
> The file is for practice.
> I hope it works!

Save the file as practicefile.txt in the same location as Python and in a different location.

Write Python code to read the file from the same location as Python and from a different location.

The read function in Python, by default, will read the whole file. It is possible to read a specified number of characters in the file or a specified number of lines.

To read a specified number of characters in a text file the function

```
read()
```

can be used. For example:

```
h = open("practicefile.txt", "r")

print(h.read(7))
```

will read the first 7 characters in your practice file. This means that the output will be:

```
This is
```

To read a specified number of lines in a text file the function

```
readline()
```

can be used. How many times this function is used in the code will be dependent on the number of lines that need to be read. For example:

```
h = open("practicefile.txt", "r")

print(h.readline())

print(h.readline())
```

will read the first two lines in your practice file. This means that the output will be:

```
This is my practice file.

The file is for practice.
```

It is also possible to loop through the file so that the file can be read line by line. This code would be written as:

```
h = open("practicefile.txt", "r")

for x in h:

print(x)
```

This means that the output from this would be:

```
This is my practice file.

The file is for practice.

I hope it works!
```

Closing a file

It is good practice to close files when actions have been completed. For example, the code shows the closing of the practice file when the first 7 characters have been read.

```
h = open("practicefile.txt", "r")

print(h.read(7))

h.close()
```

When data has been written to, or edited/appended, it is important that the file is saved and closed. If this does not happen then any changes made to the contents of the file will be lost and the file will not be up to date.

Creating a new file

New files need to be created. For example, a new business will need to create a stock file. Python defaults to "r" which will return an error message if the file does not exist. Each of the modes of opening files in Python will check if a file already exists or not; this is shown in Table 2.10.

The modes append and write will create a new file if it does not already exist. But, the create mode will return an error if the file does exist.

To create a new file the open() notation is used. The mode required is then denoted by the relevant parameter.

If a new file called "mytestfile.txt" is to be created, then the code would be:

```
h = open("mytestfile.txt", "x")
```

This expression will create a file called "mytestfile.txt" but if the file already exists an error message will be returned. If this happens then the new file name will need to be changed.

The expression to create a new file called "mytestfile.txt", if the user is sure this file name does not exist, is:

```
h = open("mytestfile.txt", "w")
```

Writing to a file

There are many occasions when an existing file will need to be appended or edited. It is important to use the correct file name when writing a file. If the incorrect file name is used, then this will result in the incorrect file's contents being changed.

To write to an existing file the open() notation is used. The mode required is denoted by the relevant parameter. The parameters are shown in Table 2.10.

The "a" parameter is used to append the contents of a file. This parameter will also create a new file if the file name does not already exist. To append an existing file's, in this case practicefile.txt, the code would be:

```
h = open("practicefile.txt", "a")

h.write("Another line has been
added.")

h.close()

# Open and read the file after the
appending:

h = open("practicefile.txt ", "r")

print(h.read())
```

The code uses a comment to explain the code purpose. This is defined with a "#" at the start of the comment.

> Comments are covered in section 2.7.1 of this content area.

The output of the code will be:

```
This is my practice file.

The file is for practice.

I hope it works!

Another line has been added.
```

As well as editing the contents of a file, it is possible to overwrite the contents with new contents.

The "w" parameter is used to overwrite the contents of a file. This parameter will also create a new file

if the file name does not already exist. To write to an existing file, in this case practicefile.txt, the code would be:

```
h = open("practicefile.txt", "w")

h.write("The contents of the file
have been totally changed.")

h.close()

# Open and read the file after the
changes:

h = open("practicefile.txt ", "r")

print(h.read())
```

The output of the code will be:

```
The contents of the file have been
totally changed.
```

The original contents of the practicefile.txt have been overwritten. So, instead of having four lines of text as output, there is now only one line.

> **Test yourself**
>
> 1 What is the purpose of the Python parameter "w"?
> 2 Write Python code to read the first 12 characters in a file named "TestYourself.txt".
> 3 What is the difference between writing a file and appending a file in Python?
> 4 Write the Python code to add "This is question 4" to the file named "TestYourself.txt".
> 5 Which parameter is used to create a new file?

2.4 Program structure

2.4.1 Using sequence, selection (branching) and iteration within programs and algorithms

An algorithm is a set of step-by-step instructions used to solve problems. There are three constructs to use when designing an algorithm:

- sequencing
- selection
- iteration.

These are the required building blocks which help to describe solutions in a format that is ready for programming. It is important to remember that algorithms contain steps and programs contain statements.

Sequence

Sequences are the main logical structure of algorithms and programs. The sequence of actions must be appropriate and run in the order that they are presented. No instruction can be skipped. There is no restriction to the number of instructions within a sequence as long as the rules are followed. Consider the actions you take when you wake up in the morning and you get up to go to school or college. An algorithm can be represented as flowcharts or pseudocode. Sequences follow a 'sequence' of actions in a specific order.

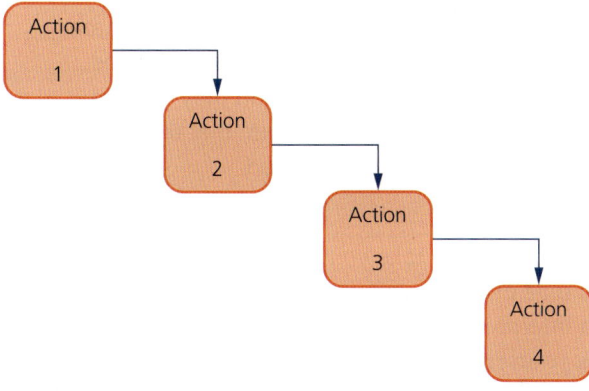

▲ Figure 2.1 A sequence of actions for an algorithm

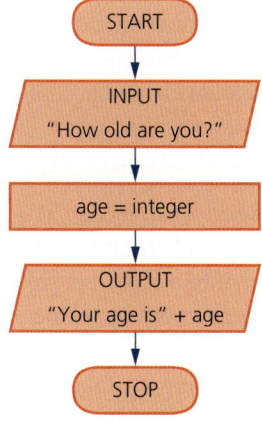

▲ Figure 2.2 Example of a flowchart to represent an algorithm using a sequence

Activity

Create pseudocode for:

The program asks the user "How old are you?"

When the user inserts their age, the program displays "You age is" + age (where + age is the age input by the user).

Selection (branching)

A selection is used when a program needs to ask a question or make a decision, because it has reached a stage where there is more than one option available. Therefore, depending on the answer/decision provided, the program will follow a particular step and ignore the other steps available. Selections are important as they allow there to be more than one path to follow within a program. Without the option to use selections within a program, there would not be the potential to take different paths. This would result in the created solutions not being realistic.

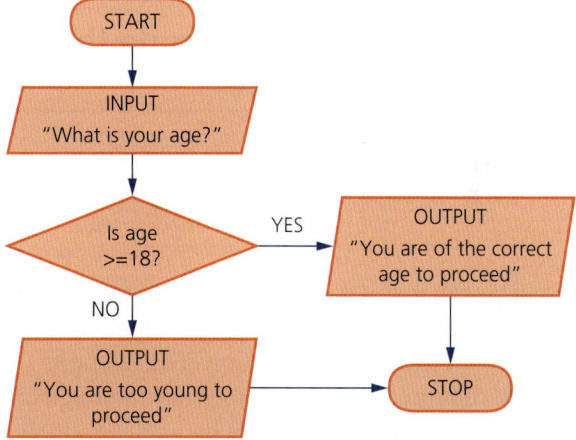

▲ Figure 2.3 A flowchart for the selection process

Iteration

Iteration is the process where a set of instructions are repeated in a sequence a set number of times or until a predetermined condition is met. When a set of instructions are completed and then executed again, this is called an iteration. Iterations simplify algorithms by stating that predetermined steps are repeated as required.

A simple example of an iteration

A person is having cereal for breakfast. This is the algorithm that could be created with the iterative steps for eating the cereal.

1 Remove cereal bowl from the cupboard.
2 Remove spoon from the cutlery drawer.
3 Put cereal in the cereal bowl.
4 Pour milk onto the cereal in the cereal bowl.
5 Use spoon to pick up cereal from bowl.
6 Put spoon in mouth.
7 Use mouth to remove cereal from spoon.
8 Chew cereal and swallow.
9 Repeat steps 5 to 8 until all cereal and milk in cereal bowl has been eaten.
10 Wash up empty cereal bowl and spoon.

Test yourself

1 Give an example of a simple iteration (without using the examples provided in this section.)
2 Identify the three constructs used when designing an algorithm.
3 Describe how a sequence works.
4 What is an iteration?
5 Use pseudocode to represent an algorithm for a customer setting up an online account:

> A customer is setting up their online account. Their email address is their username and they have to create a password. The password must contain at least one capital letter, one number and a special character. If the customer does not create a password complying with these rules, they receive an error message informing them that the password is not acceptable. If they create a password that complies with the rules, their details are stored and they receive a message informing them that their account has been created.

2.4.2 Write and interpret code that makes use of sequence

As previously stated, sequencing is the order in which the program code (statements) is executed. If the sequence is not in the correct order, then the program will not function as intended. So it is important that you have a good understanding of what the program is for (the purpose) and the sequence that the code must follow. For example, you write code to ask someone to input their age and print them out a message that includes their age. If you wrote a sequence that called for the action of printing the age, before it carried out the action of asking for the users age, it will not work. How can it display the message and the user's age before it even knows what it is? This is where pseudocode is useful as it helps you to plan the sequence of actions that needs to take place. So there are two key points to remember:

▶ What is the purpose of the sequence (what do you want to happen)?
▶ What is the sequence (the steps of the process to follow) to achieve the final output?

Having the instructions in the wrong order is a common mistake by programmers, regardless of the programming language they are using!

It is important to take your time and make sure that you read every line of your code carefully so that you can interpret what sequence of actions are taking place. Ask yourself:

▶ Are the actions in a logical order? This means that the code that you have written is in the correct order for performing calculations, carrying out processes etc.
▶ Has the code been written in the most efficient and effective way? Efficient usually means that the code is 'correct' and follows standards and accepted approaches when writing sequences. Effective means that it will achieve the desired outcome(s).

Even when taking extreme care when writing code, it is easy to make a mistake. If professional programmers can make mistakes. Python has an interactive interpreter (IDLE). IDLE is the Integrated Development Learning Environment which is included in the installation of Python. The IDLE is useful for identifying mistakes such as syntax and runtime errors. There is however another type of error referred to as a logic error. This occurs when a program doesn't do as intended even though it appears to be a valid program. These logic errors can cause unexpected results referred to as bugs. When you identify and rectify these bugs it is known as debugging. To help you debug a program, you use a tool known as a debugger. This is a useful tool that helps you identify the bugs and understand why they have occurred.

Debugging is a skill that you will use constantly when coding. Within the IDLE there is a Debug Control window that you can open to help you debug your code.

Using a debugger can be time consuming and not easy to use. Not all programming languages have debuggers available, especially for small devices used for the Internet of Things. Very few of these IoE devices have built-in debuggers. You can use something called print debugging using print(). This will display text in the console and indicates the point at where the program code is executing and the condition of the variables within the program at certain points throughout the code.

2.4.3 Write and interpret code that uses selection (branching)

In section 2.4.1 you learned what selection was and how you could use a flowchart as part of the design prior to producing the code. As a reminder, selection is to support decision making. The path taken by a program will be based on the 'answer' given to the questions asked. It is also important to notice the indentation that is used when writing the code. There are various statements that can be used in isolation or in combination. Each of the statements will be explained with an example of how they can be used. You will see how the statements can be combined to build up code to address a specific scenario:

> A thermostat is used to control the temperature within a building. If the temperature rises above say 20°C, then a message will be displayed to turn the thermostat down to 20°C. If the temperature is at 20°C, then the message temperature 20°C will be displayed. If the temperature is less than 20°C, then a message will be displayed to turn the thermostat up to 20°C.

IF

When IF is used, the program will evaluate the expression and if the expression is 'true', the statement(s) will be executed. If the expression is 'false', the statement(s) is not executed by the program.

ELIF (ELSEIF)

This keyword is basically saying 'If the previous condition(s) are "false", then test this condition.'

Example

```
# If the temperature is above
20°C, a message will be displayed
to turn the thermostat down
```

```
# If the temperature is less than
20°C, a message will be displayed
to turn the thermostat up

if   temperature > 20 :

        print(temperature, "turn
        thermostat down to 20°C" )

elif temperature < 20 :

        print(temperature, "turn
        thermostat up to 20°C" )
```

ELSE

ELSE is used to address anything which is not supported by the previous conditions.

Example

```
# If the temperature is above
20°C, a message will be displayed
to turn the thermostat down
```

```
# Or if the temperature is less
than 20°C, a message will be
displayed to turn the thermostat
up
```

```
# Else a message will be
displayed saying the temperature
is 20°C

if   temperature > 20 :

        print(temperature, "turn
        thermostat down to 20°C" )

elif  temperature < 20 :

        print(temperature, "turn
        thermostat up to 20°C" )

else :

        print("temperature,20°C" )
```

CASE

Case statements are used instead of using if-elif statements when a program code contains multiple choices. An example would be where the users are presented with a large number of different options. A case (sometimes referred to as a switch statement, in Python the match case statement) is basically a selection control mechanism and is used to allow the value of a variable or an expression to change the control of the flow of the program execution. In early versions of Python (before 3.10) there was no case statement and programmers used **dictionary** mapping instead.

Key terms

Dictionary: an iterable data structure that is built into Python. A Python dictionary has a series of 'keys' that have 'values' or 'data', so the dictionary is a set of mappings from specific keys to specific values. Unlike a dictionary we use for checking our spelling, a Python dictionary is unordered. In other words, the items in the dictionary are not in a set order. You can create an empty dictionary or a dictionary containing keys and values. Keys can be added, deleted and amended within the dictionary.

Syntax: a general set of rules for how words and sentences should be structured. These rules are known as the language syntax. When writing programming code, the syntax defines how declarations, functions, commands and other types of statements are arranged.

When writing code, it is important to test it iteratively as well as on completion. Testing is covered in detail later in this content area.

Refer to section 2.8 of this content area.

Test yourself

1 Explain the purpose of 'case' statements.
2 Describe the term 'dictionary' in relation to the Python programming language.
3 What did programmers use in Python programming for the case statement before the match case statement was introduced?
4 Draw a flowchart to show how this program would function:

   An oven has to heat up to 180°. When it has reached the correct temperature, it stops heating up and the oven light goes off.

5 Write code for the scenario described in Question 4. You should make use of a selection in your program.

2.4.4 Write and interpret code that makes use of iteration

In section 2.4.1 you learned that an iteration is where code is written to repeat certain actions either a predefined number of times (definite iteration) or until a certain condition has been met (indefinite iteration).

Loop

A loop is a structure or block of code that is used to repeat a sequence of instructions until a specific condition is met. Loops can be used for several things including adding numbers, repeating functions and cycling through values. Loops can be used in all modern programming languages but how they are implemented and the **syntax** used can be different. The most common forms of loops are the 'for loop' and the 'while loop'.

For loop

This is used for iterating a sequence that is a list, a tuple, a dictionary, a set or a string. In Python, the keyword 'for' works as an iterator method in the same way as other object-oriented programming languages. The 'for loop' does not require you to set an indexing variable first.

Example 1

```python
# Print each subject as a subject list

subjects = [ "Maths", "English", "Science", "Information Technology", "French" ]

for x in subjects :

    print( x )
```

Output

```
Maths

English

Science

Information Technology

French
```

Example 2

```
# This is how you loop through a
string

for x in " TLevels" :

    print( x )
```

The break statement

Output

```
T
l
e
v
e
l
s
```

The one important thing to remember when creating a loop is that you need to be able to force it to stop. In Python you can use the 'break' statement to stop the loop for some predefined condition.

The break statement in a 'for' loop

Example 1

```
# This is how to exit the loop
when x is Science

subjects = [ "Maths", "English",
"Science", "Information
Technology", "French" ]

for x in subjects :

    if x == "Science" :

        break
    print( x )
```

Output

```
Maths

English

Science
```

Example 2

```
# This is how to force the break
before Science

subjects = [ "Maths", "English",
"Science", "Information
Technology", "French" ]

for x in subjects :

    if x == "Science" :

        continue

    print( x )
```

Output

```
Maths

English
```

The continue statement in a 'for' loop

If you stop a loop by using the 'break' statement, you can use the 'continue' statement to continue with the next iteration in the loop.

Example 1

```
# This is how to continue with
the next iteration in the loop
without printing out Science

subjects = [ "Maths", "English",
"Science", "Information
Technology", "French" ]

for x in subjects :

    if x == "Science" :

        continue

    print( x )
```

Output

```
Maths

English

Information Technology

French
```

29

The range( ) function in a 'for' loop

When you want to loop through a set of code for a predetermined number of times, you use the range ( ) function. The range function starts with the default value of 0 and counts upwards incrementally by 1 until it finishes at a predetermined number.

Example 1

```
# How to use the range function
with a 'for' loop

for x in range(4) :

    print(x)
```

Else keyword example

Output

```
0

1

2

3
```

Although the range( ) function has a default starting value of 0, you can specify the starting value, for example if you want a starting value of 3 and an end value of 9 you would write the function as **range**(3, 9). The output would be, 3, 4, 5, 6, 7, 8.

As stated earlier, the default increment for the range function is 1, but you can specify the increment value within the range function, for example the starting value is 2, the end value is 24 and the increment value is 4. The range function would be written as **range**(2, 24, 4). The output would be 2, 6, 10, 14, 18, 22.

The else keyword in a 'for' loop

This is used to execute a specific block of code once the loop is completed.

Example

```
# Print each subject as a subject
list and print a message when
completed

subjects = ["Maths", "English",
"Science", "Information
Technology", "French"]
```

```
for x in subjects :

    print( x )

else :

    print("I'm exhausted from
    studying all of these
    subjects:")
```

Output

```
Maths

English

Science

Information Technology

French

I'm exhausted studying all of
these subjects
```

It is important to note that the else block will not be executed if the loop is stopped using a break statement.

A loop within a loop (nested loops)

The 'inner loop' will be executed each time an iteration takes place for the 'outer loop'.

Example

```
#  How to create a nested loop

marks = [95, 82]

subjects = [ "Maths", "English" ]

for x in marks :

    for y in subjects :

        print( x, y )
```

Output

```
95 Maths

95 English

82 Maths

82 English
```

The pass statement in a 'for' loop

Another important thing to note is that 'for' loops cannot be empty. If there is an empty 'for' loop, you can use the 'pass' statement to prevent an error. An empty 'for' loop is used to delay the execution of the remaining code and when a statement is required but no action is to be taken.

```
# How to prevent an error when a
'for' loop is empty

for x in [0, 1, 2, 3, 4, 5, 6] :

        pass
```

If you run this code, nothing will be displayed as it is purely to prevent an error.

While loops

The 'while' loop is used to execute a set of statements as long as a predefined condition is true. When using the 'while' loop you must prepare the variables. In the example, x has been set to 1.

Example

```
# Print x as long as x is less
than 10

x = 1

while x < 10 :

        print(x)

        x += 1
```

Output

1

2

3

4

5

6

7

8

9

The break statement

As with the 'for' loop, the break statement can be used to force the loop to stop, even if the condition is true.

Example

```
# How to exit the loop when x = 5

x = 1

while x < 10 :

        print( x )

        if x == 5 :

                break

        x += 1
```

Output

1

2

3

4

5

The continue statement

The continue statement can also be used with the 'while' loop. Remember that this enables you to stop the current iteration and continue with the next iteration.

Example

```
# How to use the continue
statement to continue to the next
iteration

x = 0

while x < 10 :

        x += 1

        if x == 5 :

                continue

        print(x)
```

Output

1

2

3

4

6

7

8

9

10

Did you notice that the number 5 is missing?

The else statement

The else statement can also be used with the 'while' loop to run a specific block of code when the condition is no longer true. In other words, when the condition is false. You would not use else after 'for' and 'while' because once the loop has completed it will automatically execute the next instruction.

Example

```
# Using else to print a message
when the condition is false

x = 1

while x < 10 :

    print( x )

    x += 1

else :

    print("x is now equal to 10")
```

Output

1

2

3

4

5

6

7

8

9

```
x is now equal to 10
```

2.4.2 Debug code that makes use of sequence

2.4.3 Debug code that makes use of selection (branching)

2.4.4 Debug code that makes use of iteration

When code does not function or does not produce the expected results, the code should be checked for errors. This is called debugging.

If the code will not run and produces an error message, it is most likely this is caused by a syntax error, which is an error in the code itself. This can be missing punctuation, typing errors in the commands or, in Python, missing colons or indentations. There will be a message to indicate what the error is and where it is.

In this case, check the commands for typos and check that the syntax is used correctly. For example, in Python, look for colons after if, elif and else statements, and check that the lines to be executed for a loop or after a conditional statement are indented.

```python
value = int(input("Enter a number
between 5 and 10 "))

if value < 5

print("value too small")

if value > 10

prnit("value too big")
```

In this example:
- the colons are missing after the if statements
- the line to be executed if the conditions returns true are not indented
- the print command on the last line is misspelled.

The code should be:
```python
value = int(input("Enter a number
between 5 and 10 "))

if value < 5 :

    print("value too small")

if value > 10 :

    print("value too big")
```

If the program runs but does not produce the expected results, then it is a logic error.

Logic errors in a sequential program will most likely be in the sequence of commands. For example:
```python
x = 4

x = x * 3

x = x + 5

    print(x)
```

will return:
```python
x = 17 ie 3*4 + 5
```

The example:
```python
x = 4

x = x + 5

x = x * 3

    print(x)
```

will return:
```python
x = 27 ie 3*(4+5)
```

Errors in if, elif and else statements that are correctly formed will most likely be caused by the incorrect choice of Boolean condition. For example, if the first condition is incorrectly set to temperature < 20, the code:
```python
if temperature < 20 :

    print(temperature, "turn
    thermostat down to 20°C")

elif temperature < 20 :

    print(temperature, "turn
    thermostat down to 20°C")

else :

    print("temperature is 20°C")
```

will return temperature is 20°C for all values of temperature that are 20 or above. This would be identified through testing using a range of typical values around the target value such as 19, 20, 21.

Errors in iterative code will usually be related to the loop conditions. In Python, the range command defaults to a starting value of 0 if not specified. To print out a 7 times table from 1 to 5, the code would not produce the desired result:
```python
for x in range(5) :

    print(x * 7)
```

It would output:
```
0

7

14

21

28
```

To achieve the desired output, we need to specify the starting value and the end value, 1 greater than the desired end value:
```python
for x in range(1,6) :

    print(x * 7)
```

With loops we also need to be careful not to specify an end point that can never be reached, as this would create an infinite loop that would not stop until the user escaped the program. For example:

```
x = 1

while x < 6 :

    print(x * 5)

x += 1
```

This program will print 5 continuously until the user kills the program. The last line has not been indented so within the loop x never changes. The code should be:

```
x = 1

while x < 6 :

    print(x * 5)

    x += 1
```

Another common problem is using values that can result in division by zero. The program will crash if this happens.

The program is intending to divide 100 by values of x from 5 down to 1, but the loop condition is not set correctly and the program crashes once x becomes 0.

```
x = 5

while x >= 0 :

    print(100 / x)

    x = x - 1
```

The condition should be:

```
x > 0
```

2.4.5 Declare and call functions and procedures

A function is code which is written to perform a specified task and it can be used more than once within a program. Functions contain a set of instructions in a subroutine and are called when required. Programming languages provide numerous inbuilt functions, for example in Python you have print( ), input( ), exec( ), compile( ), help( ) and so on. It is also possible to create your own functions. In order to carry out a function, it may need multiple inputs.

Python has three different types of functions:

- ▶ **built-in functions** – as previously stated, Python provides numerous built-in functions
- ▶ **user-defined functions (UDFs)** – functions that can be created by software developers
- ▶ **anonymous functions** – sometimes referred to as lambda functions because they are not declared using the standard 'def' keyword.

Methods versus functions

A method is a function which is part of a class and accessed using an instance or an object of the class. A function is not restricted and is standalone. Therefore, all methods are functions, but not all functions are methods.

How to create a simple function

In Python a function is defined using the keyword 'def'.

Example

```
# How to create a simple function

def my_simple_function( ) :

    print( "This is my simple function" )
```

How to call a function

It is very simple to call a function. You use the name of the function followed by parentheses.

Example

```
my_simple_function()
```

Output

```
This is my simple function
```

You do not have to include print( ) when you call the function because it is already in the function that you have created.

Arguments and parameters

An **argument** is the value that is sent to the function when it is called. A **parameter** is the variable that is listed inside the parentheses in the function definition.

Example 1

```
# List of friends

def my_simple_function( friends ):

    print(friends, "good friend" )
```

```
my_simple_function( "Saundra" )

my_simple_function( "Sonia" )

my_simple_function( "Touffouo" )
```

Output

```
Saundra     good friend

Sonia       good friend

Touffouo    good friend
```

The argument is "friends". You can have more than one argument but you must separate them with a comma. Therefore, if a function is anticipating three arguments, then when calling the function you must also pass all of the arguments. In Python documentation, arguments is usually shortened to args.

Example 2

```
# A function with two arguments

def my_function( fname, lname):

    print( fname + "  " + lname )

my_function( "Touffouo" ,
"Mahnou" )
```

Output

```
Touffouo Mahnou
```

If you get an error when running a function with multiple arguments, check that you are calling all of the arguments within the function. This is a common mistake people make.

Procedures

Every procedure requires:
▶ a name
▶ the programming code to perform the required task.

A function is compiled before it can be called and can be called multiple times within a program. A procedure is only compiled once. A function returns a value and control the the code of function that is calling it. A procedure returns the control but does not always return a value to the code or function that is calling it.

Example

Let's return to the thermostat scenario in section 2.4.3. First of all the procedure needs to be given a name. So, we will call it thermostat control.

```
def    thermostat_control( ) :
```

The procedure code is then written underneath but notice that it is indented.

```
def thermostat_control (temp) :

    if temp > 20 :

        print(temp, "turn
        thermostat down to 20°C" )

    elif temp < 20 :

        print(temp,"turn thermostat
        up to 20°C" )

    else:

        print("temperature is 20°C")
```

Indenting the code in this way informs Python that the code belongs to a procedure.

Running a procedure

A procedure can be called at any point in a program and can be called multiple times. When using Python, a procedure is called by using its name including the brackets. Looking at the thermostat example, the procedure to control the temperature could be called at regular intervals, for example at the end of every hour to check the temperature and, depending on the temperature recorded, it would send out the relevant message. Do not confuse functions and procedures. A function will always return a value, whereas a procedure need not return a value.

Therefore the procedure could be called every hour with the current temperature and would send out the relevant message.

The procedure would be called by inserting:
```
thermostat_control(<current
temperature>)
```

for example
```
thermostat_control(19)
```

2.4.6 How standard searching and sorting algorithms work, and the benefits and drawbacks

Searching algorithms

Linear search

This is the simplest method of searching a data set and is sometimes referred to as a 'serial' search. Starting at the beginning of the data set, each item of data is examined until a match is made. Once the item is found, the search ends. If there is no match, then the algorithm must be able to deal with this. An example would be looking up a telephone number in your contacts on your mobile.

Example

Let's assume that you want to carry out a linear search in an array for the letter F.

Index	0	1	2	3	4	5	6	7	8
Data	A	D	G	C	B	E	H	F	I

The search will start at index 0. The data (element) at index 0 will be compared with the search criteria. The question is asked "Does F = data in index 0?" The answer is no and therefore the comparison is carried out with the data in the next index. The question is

asked "Does F = data in index 1?" Again, the answer is no, so the comparison is conducted with the data in index 2. This process is continued until either F is found (in this example it will be found at index (7) or there are no further data in the array to compare with.

Binary search

The binary search is a fast search algorithm and works on the principle of divide and conquer. In order for a binary search to work as intended, the data should be in a sorted format. The search interval is repeatedly divided in half. At the start of the process, the interval covers the entire array. If the value of the middle item in the interval is greater than the value of the search key, then the interval is changed to the lower half of the array. If the search value is lower than the middle of the array interval, then the interval is changed to the upper half of the array. This process continues until the value is either found or the interval is empty.

Example

Imagine we have a list sorted alphabetically with 9 items and we wish to locate the letter F.

A D F G H J K L M

The mid-point of the list is the letter H.

F is before H in the alphabet, so we can discard all the items from the letter H to the end of the list.

This leaves a new list: A D F G.

The mid-point of this list is D, as when finding the mid-point of an even number of values, we generally choose the value to the left of the middle.

F follows D alphabetically so we can discard the lower part of the list up to the letter D.

This leaves a new list: F G. The mid-point is the letter F. We have found the value.

Benefits and drawbacks when using searching algorithms

Benefits	Drawbacks
Linear search	
• The list does not have to be ordered • Fast performance when searching small to medium lists • Because the list does not have to be sorted (ordered), additional elements can be added and deleted. When searching, algorithms that require an ordered list will mean that any additions or deletions require the list to be re-sorted. This means that a linear search is more efficient • The linear search is simpler and therefore it is easier to write program code to carry it out • It can carry out a search on any type of data	• It is very slow when searching lists which have vast quantities of data items
Binary search	
• Faster performance than a linear search because the data that requires searching is halved at each stage	• Data must be sorted (ordered) first

▲ Table 2.11 The benefits and drawbacks of linear and binary searches

Sorting algorithms

Bubble sort

The bubble sort repeatedly swaps adjacent elements that are not in order, until the entire list of items is in sequence. The items "bubble up" the list according to their key values.

Example

Sort the array into ascending order (57, 32, 83, 28, 7)

First pass

The first two elements are compared (is 57 < 32). The answer is no, so the two values are swapped. The array then has the order of (32, 57, 83, 28, 7).

Elements two and three are then compared (is 57 < 83). The answer is yes, so the two values stay as they are. So the array remains in the order of (32, 57, 83, 28, 7).

Elements three and four are now compared (is 83 < 28). The answer is no, so the two values are swapped, and the order of the array becomes (32, 57, 28, 83, 7)

Elements four and five are compared (is 83 < 7). The answer again is no, so the two values are swapped. The order of the array becomes (32, 57, 28, 7, 83)

Second pass

The first two elements of the array are compared (is 32 < 57). The answer is yes so, the order of the array remains unchanged.

Elements two and three are compared (is 57< 28). The answer is no, so the two values are swapped, and the order of the array becomes (32, 28, 57, 7, 83)

Elements three and four are compared (is 57 < 7). The answer is no, so the two values are swapped, and the order of the array becomes (32, 28, 7, 57, 83).

Elements four and five are compared (is 57 < 83). The answer is yes and so the order of the array remains unchanged.

Third pass

The first two elements of the array are compared (is 32 < 28). The answer is no and so the two values are swapped. The order of the array becomes (28, 32, 7, 57, 83).

The second and third elements of the array are compared (is 32 < 7). The answer is no, and the two values are swapped. The order of the array becomes (28, 7, 32, 57, 83)

The third and fourth elements are compared (is 32 < 57). The answer is yes and so the order of the array remains the same.

The fourth and fifth elements are compared (is 57 < 83). The answer is yes and again the order of the array remains unchanged.

Fourth pass

The first two elements are compared (is 28 < 7). The answer is no and so the two values are swapped. The order of the array becomes (7, 28, 32, 57, 83).

The second and third elements are compared (is 28 < 32). The answer is yes and so the order of the array remains unchanged.

The third and fourth elements are compared (is 32 < 57). The answer is yes and so the order of the array remains unchanged.

The fourth and fifth elements are compared (is 57 < 83). The answer is yes and so the order of the array remains unchanged.

Fifth pass

The first and second elements of the array are compared (is 7 < 28). The answer is yes and so the order of the array remains unchanged.

The second and third elements of the array are compared (is 28 < 32). The answer is yes, and the order of the array remains unchanged.

The third and fourth elements are compared (is 32 < 57). The answer is again yes and therefore the order of the array remains unchanged.

The fourth and fifth elements are compared (is 57 < 83). The answer is yes, and the order of the array remains unchanged.

The algorithm has now confirmed that there are no further changes required.

Insertion sort

The insertion sort repeatedly scans the items in the list. Each time it scans, it inserts the item into its correct position.

Example

Imagine a list of shoe sizes 5 to 12 but they are not placed in the correct order, for example, 11, 7, 10, 5, 8, 12, 6, 9. The insertion sort method can be used to place the shoe sizes in the correct order.

First scan

11 > 7 therefore the two values are swapped. The resulting order is:

7, 11, 10, 5, 8, 12, 6, 9

Second scan

7 < 11 (so it remains where it is).

11 > 10 so the two numbers are swapped. The resulting order is:

7, 10, 11, 5, 8, 12, 6, 9

Third scan

7, 10 and 11 are in the correct ascending order so remain where they are.

11 > 5 so 5 is placed as the beginning of the list as it is smaller than 7, 10 and 11. The resulting order is:

5, 7, 10, 11, 8, 12, 6, 9

Fourth scan

During this scan, it will identify that 8 is in the wrong place as 11 > 8, then 8 will be moved to the position after 7 (because it will check each of the values before 11 to see whether the value is higher or lower than 8). The resulting order is:

5, 7, 8, 10, 11, 12, 6, 9

Fifth scan

During the fifth scan, it will identify that the next value not in the correct order is 6 as 12 > 6. It will compare the value of 6 with each of the preceding values before 12 and place 6 in the correct position. The resulting order is:

5, 6, 7, 8, 10, 11, 12, 9

Sixth scan

During the sixth scan, it will identify that the next value not in the correct order is 9 as 12 > 9 (all preceding values are in the correct order). The value 9 will be compared with each of the proceeding values and placed in the correct position. The resulting order is:

5, 6, 7, 8, 9, 10, 11, 12

Merge sort

This uses a technique referred to as divide and conquer. It is a more complex form of sort but also very efficient. The list is repeatedly divided in two until all the elements are separated individually. Pairs of elements are compared, placed in order and combined. The process is repeated until the list is recompiled as a full list again.

Selection sort

(Note: This is not required for the exam, but will be useful for you to know.) This works by repeatedly going through the list of items and, each time, selecting an item according to its ordering. It is then placed in the correct position in the sequence.

Example

Consider the array:

| 15 | 24 | 27 | 8 | 19 | 7 |

The first position in the array is identified (15). The rest of the array is scanned to identify the lowest number (which is 7).

| 15 | 24 | 27 | 8 | 19 | 7 |

So the two elements are swapped.

| 7 | 24 | 27 | 8 | 19 | 15 |

The second position is found (24) and the lowest value is found after scanning the rest of the array (8).

| 7 | 24 | 27 | 8 | 19 | 15 |

The two elements are swapped.

| 7 | 8 | 27 | 24 | 19 | 15 |

The process continues as follows:

7	8	27	24	19	15
7	8	15	24	19	27
7	8	15	24	19	27
7	8	15	19	24	27

The selection sort is now complete.

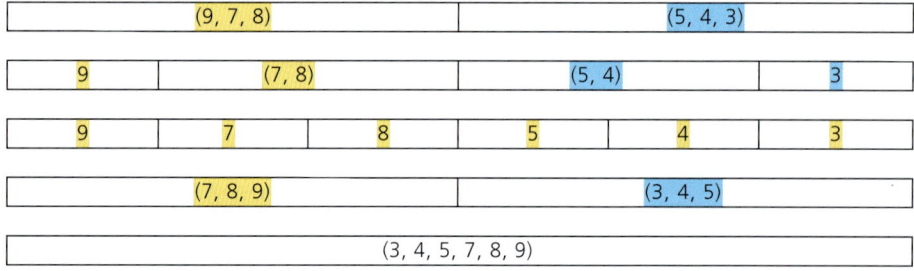

▲ **Figure 2.4** An example of the merge sort process

Quick sort

The quick sort works on the same principle as the divide and conquer principle used in the merge sort. The list of items is partitioned into two sub-lists based on a pivot element. All of the elements in the first sub-list are arranged so that they are smaller than the pivot. All of the elements in the second sub-list are arranged to be larger than the pivot. This process is repeated on the resulting sub-lists until all of the items are sorted.

Benefits and drawbacks of sorting algorithms

Benefits	Drawbacks
Bubble sort	
• Easy to implement • Elements are swapped in place as opposed to using additional storage • Minimum space requirements	• Does not work as well with a list containing vast numbers of items • Requires n-squared processing steps for every n number of elements to be sorted • Not suitable for real-life applications
Insertion sort	
• Simple to use • Has good performance when sorting a small list • Minimal space requirements because it is an in-place sorting algorithm	• Does not perform as well as other sorting algorithms • Requires n-squared processing steps for every n number of elements to be sorted, and therefore does not perform well with a list with a vast number of items • Only really useful when sorting a list containing just a few items
Merge sort	
• Can be used for files of any size • The reading of the input during the run-creation step is sequential • As the sort is used for the in-memory section of the merge, its operation can be overlapped with input/output (I/O)	• Requires extra space • Requires more space than other forms of sort • Less efficient than other forms of sort
Selection sort	
• Performs well on a small list • An in-place sorting algorithm therefore no additional temporary storage required except what is required to store the original list • Performance affected by the initial ordering of the items prior to the sorting process	• Not very efficient on a list with a vast number of items • Requires n-squared number of steps for sorting n elements • Not as efficient as quick sort
Quick sort	
• Considered to be the best sorting algorithm • Good for dealing with lists with a vast number of items • Sorts in-place and therefore requires no additional storage	• Can have similar average performance as bubble, insertion and selection sorts • If the list is already sorted, it is not as efficient as bubble sort • If the sorting elements are integers then it is not as efficient as other forms of sort

▲ Table 2.12 The benefits and drawbacks of sorting algorithms

Test yourself

1 Describe the term 'bubble sort'.
2 Show how merge sort would process the array:

21	17	33	15	42	7	27

3 You have a list that contains a huge number of items. Select the best sorting method with respect to speed and effectiveness.
4 Sort the array into ascending order (9, 4, 7 3, 6, 2) using the bubble sort.
5 An array contains numerical values from 1 to 75. Show how a binary search would find the value of 30.

2.5 Built-in functions

2.5.1 The benefits and drawbacks of using pre-written code

Pre-written code is code that already exists when you are developing new software. It may be within your own organisation or from an external source. The purpose of using pre-written code may be to carry out the same function (or something similar). There are a number of ways that code can be reused including:

▶ using third party libraries
▶ using open-sourced frameworks
▶ repurposing code that has been used/developed/ written internally.

There are a number of benefits and drawbacks to using pre-written code.

Benefits

▶ **Shorter development time** – the Python Package Index is an example of a tool that helps developers find pre-existing code to add functionality to the software they are developing. This reduces the time needed to develop a solution as the developers do not have to plan and write the code as it is readily available.
▶ **Improved end product** – the use of pre-existing code enables developers to add greater functionality to a program and therefore create a better product.

Drawbacks

▶ **Security risks** – one of the biggest drawbacks is the security risks that can be an issue when using

code from third parties (and even code produced internally). Code must always be tested for security vulnerabilities; if it has not been tested, then the security of the entire program is at risk.

Industry tip

Watch this video about the vulnerabilities that can be planted in open-source code:

https://www.zdnet.com/video/open-source-planted-vulnerabilities-create-big-problems-for-developers/

It emphasises the seriousness of using source code which has not been tested for vulnerabilities prior to its use.

Another example is the software data breach caused by the Apache Commons Collection which is a third-party component for Java frameworks. Version 3.2.1 has a security vulnerability that allows an attacker to control the server. It is important that software developers are aware of these types of threats so that they can ensure that they either use a later version where the vulnerabilities have been addressed or implement other security measures. Whenever a developer considers the use of pre-written code, they need to carry out security tests to ensure that they are not exposing their own software to data breaches, **injection flaws**, **cross-site scripting** and so on.

Key terms

Open source: software where the copyright holder grants users the rights to use, edit and distribute the source code.

Injection flaws: allow attackers to relay malicious code through an application to another system.

Cross-site scripting (XSS): usually found in websites and/or web applications that accept end-user input. This can include search engines, login forms, comment boxes and message boards.

Industry tip

Small beginnings can create larger problems – insecure code used in just a small part of a program can create large-scale problems when used within a larger program and/or system.

▶ **Lack of control** – if changes are made to the code (because you do not own it), the changes may have an impact on the program being developed. A developer would need to make immediate changes.

▶ **Naming conventions** – the naming conventions used within the pre-written code and even the naming of the code can be different to that of the programming language a programmer is using to develop software. If you refer to section 2.7, you will see Table 2.13 which highlights the naming conventions used within Python.

Test yourself

1 Identify two ways that code can be reused.
2 Describe the term 'pre-written code'.
3 Explain why using pre-written code shortens development time.
4 Explain why naming conventions can be a drawback when using pre-written code.
5 Discuss the security issues that can occur when using pre-written code.

2.5.2 Select and justify the use of pre-written code provided by the Python programming language (e.g. built-in functions, standard libraries)

2.5.3 Write code that makes use of user-written and pre-written code (e.g. built-in functions, standard libraries)

As discussed previously, there are both benefits and drawbacks for incorporating pre-written code into new software that is being developed. Python has built-in functions and standard libraries which can be readily accessed.

▶ **Built-in functions** – these are functions which have already been defined in a program with a set of statements. The statements, when used in conjunction with other statements and/or code, perform tasks. This means that users do not need to create code to perform a particular function as it can be selected from the available

built-in functions. Consider the built-in functions in spreadsheets, for example =sum(A4:E4). The built-in function here is sum(). This saves time as opposed to having to input the full code of =(A4+B4+C4+D4+E4).

▶ **Standard libraries** – contain reusable sections of code (called modules) that can be incorporated into other programs/projects.

Example of using a built-in function and standard library function

You want to use the standard variables for someone to enter their date of birth. The variables will be year, month, day. You would create a date in the standard ISO format.

```
year = 1998

month = 07

day = 02
```

To create this module, you can use a function within the standard library. The function you would select is datetime module. You would then use date(year, month, day) . isoformat( ) to convert the date. In order to print the result you would use the built-in function print( ).

```
import datetime

iso_date = datetime.date( year,
month, day ).isoformat( )

print( iso_date )
```

The code could be written more succinctly as:

```
import datetime

print( datetime.date( year,
month, day ).isoformat( ) )
```

Test yourself

1 Explain the term 'built-in functions'.
2 Explain the term 'standard library'.
3 Which built-in function could you use to display a result from a module on the screen?
4 Explain the term 'module'.
5 Justify the use of built-in functions and standard libraries when developing programs.

2.6 Validation and error handling

2.6.1 Different types of input validation/Write, interpret and debug code that uses validation techniques

Input validation refers to the check made by a computer to ensure that entered data is sensible and reasonable. It is important to remember that this is not a check for the accuracy of the data.

Abnormal inputs and/or conditions are dealt with using error-handling routines in programs. It is important that error messages are clear and provide users with the options available to resolve any issues. This should not be confused with exception handling that is built into the programming language of an application or hardware in order to respond to and resolve abnormal conditions.

Commonly used validation methods are:
▶ presence check
▶ length check
▶ type check
▶ format check
▶ range check
▶ check digit.

Presence check

A presence check checks to see if a required field that should contain data has been left blank. It does not check the accuracy of the data, just that there is data available. Presence checks are usually implemented for important fields which must not be left blank, for example the credit card number to pay for goods.

The code asks the end user to insert their name and is an example of presence check validation. Enter the code into Python and, instead of inserting your name, just press enter and see what happens.

```
Is_Name_Present = False

while Is_Name_Present == False :

    ans = input("enter your
    name; " )

    if len(ans) > 0 :

        Is_Name_ Present = True

    print( "Thank you for
    entering your name")
```

Length check

This ensures that the data entered does not contain fewer or more characters than the specified minimum or maximum number. This is an important validation technique for a fixed length field being used to store data. Without this check, any additional characters that exceed the maximum number of characters would be lost.

Now let's look at the same code and change the 'len' from zero to 4.

```
Is_Name_Present = False

while Is_Name_Present == False :

    ans = input( "enter your
    name; " )

    if  len(ans) > 4 :

        Is_Name_Present = True

    print("Thank you for entering
    your name")
```

This now becomes a length check because len (which means length) is greater than zero. So it is no longer a presence check but a length check. So if someone only entered their initials, for example HE (which is less than 4 characters) instead of their name, there would be an error.

Type check

This is used to ensure that the data type that has been input is of the correct type. For example, if you want the user to enter their age and you only want the data type to be an integer, you would use a validation technique to check what the user has entered and provide a message if they have input the data incorrectly.

Example

```
#Declare the variable valid_age
and set it to false

valid_age = False

#Prompt the user to enter their
age

age1 = input("Enter your age ")

#The value entered by the user
is checked to see if it is an
integer
```

```
if age1.isdigit() :

  valid_age = True

else :

  print("Your age has not been
entered in the correct format; it
should be an integer")
```

Format check

A format check can be used to check that the data input by the user is of the correct format e.g. date, postcode, national insurance number, etc. Python also uses a technique known as try or except. In this method, it will compare the input from the user to see if it is in the correct format. If it is not, then except will capture the exception and produce an error message.

Example

```
#Import the datetime library

import datetime

#Declare the variable birth_date

birth_date = str(input("Enter
date of birth "))

try:

#Set the format you want the date
to be in

  datetime.datetime.strptime
  (birth_date,'%d-%m-%Y')

except:

  print("The date is not in the
correct format, it should be
DD-MM-YYYY")
```

Range check

This is used to check that the input value lies between a range of set values. For example, does the user input lie between the values of 45 and 72?

```
#Declare the variable target_
value and set it to false
```

```
target_value = False

while not target_value :

#Prompt the user to enter a value

  tv1 = int(input("Enter the
target value "))
    if tv1 <= 72 and tv1 >= 45 :

      target_value = True

#The message will be displayed if
the target value entered is in
the specified range

    print("The target value
is valid and the target value
entered is ",tv1)

  else:

#The message will be displayed if
the target value is not within
the specified range

    print("The target value does
not lie between 45 and 72")
```

Check digit

It is well known that errors can occur and data may be incorrectly received. To compensate for this problem, an extra value is transmitted to determine if the data is correct or incorrect. The additional value is known as a check digit.

An example of calculating a check digit on a barcode 1234567 is:

1 The first, third, fifth and seventh numbers are each multiplied by three.
2 The results from above are added together with the second, fourth and sixth numbers.
3 The total value is then divided by 10.
4 The check digit is then calculated by subtracting the remainder from 10.

Therefore:

$(1 \times 3), (3 \times 3), (5 \times 3), (7 \times 3)$

$3 + 9 + 15 + 21 + 2 + 4 + 6 = 60$

60 / 10 = 6 (there is no remainder; if, for example, the answer had been 69, there would be a remainder of 9 when divided by 10)

A barcode uses EAN codes (European Article Numbers). The calculation of the check digit is calculated using modulo 10 with a weighting of three.

In this example 10 − 0 = 10. Therefore, the check digit would be 10.

If the remainder had been 9, then 10 − 9 = 1, then the check digit would be 1.

Test yourself

1   Describe the term 'validation'.
2   Describe the term 'error handling'.
3   Explain the role of a presence check.
4   Write the Python code for a maximum length check of 15 for a variable called 'town'.
5   A barcode is in the format 2,6,4,5,1,3,7. Calculate the check digit.

2.6.2 Developing reliable and robust code

It is always important that the code developed is of a good quality. This is because the software will be more durable and easier to use and maintain. In order to be good quality, it will meet the end-user requirements and not suddenly crash.

Robust and reliable code

'Robust and reliable' means that the code can deal with errors during execution and be able to deal with the input from the end user if it isn't correct. If a program requires the end user to enter their first name and their last name, and they enter letters for the first name and accidently input a combination of numbers and letters for their last name, it is important that the program doesn't stop functioning. The program must be able to deal with this error and provide an error message to the end user so that they can make changes to the data they input. A program which is robust and reliable can handle any errors such as incorrect data input by the end user without stopping the running of the program. The addition of error messages to identify the errors helps to make the program, and therefore the code, robust and reliable.

2.7 Maintainable code

2.7.1 The accepted style conventions (such as Python's PEP8) and how these are implemented to create readable and maintainable code

Every programming language has defined style conventions. These are to ensure that the code is consistent in format and readable by users. By having code that is readable with a consistent format, the maintainability is also increased.

The current style convention for Python is PEP8. PEP stands for Python Enhancement Proposal. The PEP8 guidelines provide details relating to:
▶   naming conventions/styles
▶   use of comments
▶   use of white space
▶   the structure of the code
▶   indentation.

As Python is an open-source programming language it is not possible to enforce the guidelines defined in PEP8. However, by using the PEP8 guidelines, Python code can be created which can be read, maintained and used by others.

Research

Locate and read the 'Zen of Python' created by Tim Peters in 1999.

Naming conventions

'Explicit is better than implicit.'

The Zen of Python, Tim Peters

When creating code it is important to use logical and sensible names making it easier for someone else to work out what these represent. This can be very important when maintenance, routine or remedial, is carried out on the code. Irrelevant or inappropriate names can make it difficult to locate errors during remedial maintenance.

> Variables were covered in section 2.1.4 of this content area.

It is very important that single letter names of I, O, l are not used. This is because they can be mistaken for 1 or 0.

Table 2.13 shows the common naming conventions and styles that are used in Python and when each should be used.

Construct	Convention/style	Example expression
Constant	Capital single letter, word or words Words should be separated with underscores	X, Y CONSTANT LONG_CONSTANT
Function	Lowercase word or words Words should be separated with underscores	function long_function
Module	Short, lowercase word or words Words should be separated with underscores	module.py pay_module.py
Variable	Lowercase single letter, word or words No white space or special characters Words should be separated with underscores	x, y last_name

▲ Table 2.13 The common naming conventions and styles used in Python

Use of comments

'If the implementation is hard to explain, it's a bad idea.'

The Zen of Python, Tim Peters

Comments in any code are used to explain the function of any given section. By using comments the code becomes easier to maintain as the meaning of the code is explained.

In Python a comment always starts with a # symbol. When the code is being compiled or executed, any line in the program code that begins with the # symbol will be ignored.

This code was created to carry out mathematical operations; comments shown in red have been added to explain the purpose of the code.

```
x = 31

y = 10

z = 0
```

```
# To add two numbers together

z = x + y

print("Line 1 - value of
    z is " , z)

# To subtract two numbers

z = x " y

print("Line 2 - value of
    z is " , z)
```

Use of white space

'Sparse is better than dense.'

The Zen of Python, Tim Peters

White space can be very useful when code is being read and helps with maintenance. If there is not enough white space, then the code can look cluttered and can become difficult to read and understand. However, if there is too much white space it can be difficult to understand which elements of the code should be read together.

When using operators, mathematical, relational or Boolean, PEP8 advises that a single space should be left either side.

> Operators were covered in section 2.2 of this content area.

```
x = 31

y = 10

z = 0

# To add two numbers together

z = x + y

print("Line 1 - value of
z is " , z)
```

Looking at the example code above, it can be seen that a space has been left either side of the + operator. This makes it easier to read. If no spaces are left, the code becomes more difficult to read and understand. This can lead to possible errors when maintenance is being carried out.

```
x = 31

y = 10

z = 0

# To add two numbers together

z = x + y

print("Line 1 - value of
z is " , z)
```

When multiple operators are being used in the same expression the use of white space can become confusing. PEP8 recommends that only the lowest priority operators should have space.

Here the same expression has been provided with the correct and incorrect use of white space. Comments have been used to identify this.

```
# Correct use of white space

a = b**5 + 9

c = (a+b) * (a-b)

# Incorrect use of white space

a = b ** 5 + 9

c = (a + b) * (a - b)
```

The structure of code

The structure of code is also known as the layout. How the code is structured will have an impact on the readability and ease of maintenance.

One element of code structure is line length. PEP8 recommends a maximum line length of 79 characters, including any white spaces. Other layout elements include:

▶ indentation
▶ tabs or spaces
▶ blank lines
▶ line breaks.

Research

PEP8 provides guidelines about code structure or layout. Create a digital communication, including the relevant lines from the Zen of Python and example code, to explain how Python code should be structured.

Test yourself

1. What does 'PEP' stand for?
2. Which construct should be styled in capital letters?
3. How do comments help when code is being maintained?
4. How many spaces should be left either side of a single mathematical operator?
5. Identify one element of code layout.

2.8 Testing

2.8.1 The fundamental importance of testing for all components

Software

Software testing is a check to establish whether the software product (the program) meets the requirements and that it is defect free. Software testing involves the execution of software/system components using manual or automated tools. The purpose is to identify any errors, gaps or missing requirements when compared to the requested requirements. Software testing is the 'verification of the application under test' (AUT).

Software testing is very important, so that any bugs and/or errors in the software can be found. Testing enables any issues to be identified early and resolved before the software is deployed. If software is tested correctly, it ensures the security, reliability and performance of the product. This will save time, be cost effective and make the customers happy (customer satisfaction). Software bugs can be expensive and even dangerous in some situations. A search of the internet will provide you with many examples of where software was insufficiently tested and caused serious problems.

Benefits of software testing

▶ **Security** – people are looking for products that can be trusted. This is the most sensitive and vulnerable benefit of software testing as it ensures that risks and problems are removed early, prior to full deployment.
▶ **Cost effective** – this is an important benefit of software testing as it can save money by identifying and fixing any issues (bugs) early, therefore costing less to fix.
▶ **Product quality** – this ensures that the product meets the requirements of the intended customer(s) and therefore a quality product can be provided.

▶ **Customer satisfaction** – it is extremely important that the customers are satisfied. User interface (UI) and user experience (UX) testing ensures that the customer is satisfied with the product and that the product functions as intended.

Compatibility testing

It is important to carry out compatibility testing to ensure that the software can run on different hardware, operating systems (OSs), applications, network environments and/or mobile devices. This is commonly known as compatibility testing and is a form of software testing.

Compatibility testing is a type of non-functional testing. There are different types of compatibility tests:

▶ **Hardware** – this checks that the software is compatible with different hardware configurations. This is important because the software is of no use to the intended end user (customer/stakeholder) if they cannot run the software.

▶ **Operating systems** – this is to ensure that the software is compatible with different OSs (the customer may be using a variety of platforms).

▶ **Software** – this is to ensure that the software is compatible with other software. A program that has been developed may be required to work with other software. It is important that this is tested.

▶ **Network** – the performance of a system in a network environment is evaluated by varying the bandwidths, operating speed and capacity. It may be necessary to check that the developed program can run on different networks.

▶ **Browsers** – the developed software may need to run using different browsers such as Internet Explorer, Google Chrome and Firefox.

▶ **Devices** – this is a check to ensure that the developed software is compatible with different devices such as printers, scanners, Bluetooth, media devices and USB port devices.

▶ **Mobile** – developed software may have to be used on mobile devices (e.g. for use by remote workers). It is therefore important that any developed software is compatible with mobile devices, for example IOS and Android. If the software is not compatible, the remote workers will not be able to carry out their tasks using the developed software.

It is also important that the software is adaptive and will respond to:
a) different input methods e.g. touchcreen, keyboard
b) changes to a screen's orientation e.g. portrait/landscape
c) the type of browser used for different mobile devices

▶ **Software versions** – a system may contain various versions of pre-installed software which the developed software has to work with. It is therefore important that testing takes place to ensure that the developed software is compatible with the various versions of software that may be used alongside it.

▶ **Data** – any developed software must be tested using data that will confirm the functionality of the program. There are different forms of data testing which include not just normal data, but erroneous data, extreme data and boundary data.

There are two types of compatibility testing: backward and forward compatibility.

Backward compatibility

This testing is carried out to confirm that hardware and/or software is compatible and can function as intended on older versions of software and hardware already installed/used on systems.

Forward compatibility testing

This form of testing is to ensure that the hardware and/or software is compatible (functions) with new versions of hardware/software.

If an organisation has customers and/or offices all over the world, the hardware systems and devices can all be different. Tests must be carried out to ensure that any developed software can function as intended, regardless of where it is being used.

Resulting service (final product)

Beta testing is performed by real users of the software application in a real environment. It is one form of user acceptance testing (UAT). The beta version of the software is released to a limited number of end users of the software program to obtain feedback on the quality of the product. It is important as it helps to mitigate risks of failure and provides an increase in quality through something known as customer validation. This is the final test before the product is fully released to all intended end users. There are different forms of beta testing:

▶ **Traditional** – when the software program is released to the target customer and related data is gathered as feedback. This is used for product improvement.

▶ **Public** – where the software program is released to the world through online channels. Improvements can be made to the program based on the feedback collected. Microsoft conducted the largest beta test for its Windows 8 OS prior to officially releasing it.

▶ **Technical** – the software program is released to a group of employees in a specific organisation and feedback/data is collected and analysed.

▶ **Focused** – the software program is released to the target audience to collect feedback on specific features of the program.

▶ **Post-release** – the software is released to the intended end users and data is collected and analysed so that improvements can be made to future releases (versions) of the software.

UAT is a process that verifies that a software program (or other information technology (IT) solution) works for the end user as intended. UAT is used to assess:

▶ Can the user use the software?

▶ Does the end user have difficulty using the software?

▶ Does it meet their requirements?

▶ Does it function as expected?

UAT is sometimes referred to as beta, application and end-user testing. It is often considered the last phase in the software development process: the last test before the product is finally released to, or installed for, the customer (end user) or for final distribution to the world.

Test yourself

1 Explain the difference between forward and backward compatibility testing.
2 Describe the term 'public beta testing'.
3 Identify and describe three benefits of conducting software testing.
4 Discuss the importance of carrying out software testing.
5 An organisation has offices in different countries as well as remote workers. They have had new software developed. Identify and describe the different compatibility tests that they would need to carry out to ensure that all relevant personnel can access and use the software effectively.

2.8.2 Testing and quality assurance methodologies to seek out problems and issues

Testing was discussed in section 2.8.1. It ensures that the IT solution developed, whether that is a software program or a network solution, meets not only the end-user requirements but functions as intended. But it does not end there. There is something known as quality assurance (QA) testing. There are a number of different types of QA testing and they are all equally important.

Concept testing

Market research is used to understand the strengths, weaknesses and areas of potential improvement for a particular concept. This could be any form of IT solution, for example development of software, networks or digital systems. Interested internal/external stakeholders are provided with information relating to the basic concept. They then provide feedback which is collated and analysed for consideration prior to the product being fully developed and/or released. Concept testing can cost time and money, but the research can also save a lot of unnecessary costs in the long term.

Unit testing

This is where the individual components or units of a software program are tested. This is to ensure that each unit/component of software code functions as intended. This is carried out during the development phase of an application. Isolated sections of code are tested and its accuracy is verified. A unit may be an individual function, procedure, method, module or object. Unit testing is known as a **'white box' testing** method that is carried out by the software developer.

Key term

White box testing: a method of testing when the internal structure and design of the software is known to the tester (usually carried out by the software development team).

Integration testing

This is the process of testing the interface between two software components (units). It used to identify faults in the interaction between integrated units and is carried out after all the units have undergone unit testing. There are four types of integration testing:

▶ **Big-bang integration** – all the units are put together and tested. This approach is only suitable for very small programs and/or systems. If an error is found, it can be very difficult to identify where the error is as it could be caused by any of the units.

▶ **Bottom-up integration** – where the lower-level units are tested with the higher-level units. The system/program is broken down into subsystems with the interfaces being tested between the units that make up the subsystem.

▶ **Top-down integration** – where the low-level units that have not yet been integrated are therefore simulated. The high-level units are tested and then the low-level units are tested. The low-level units are then integrated with the high-level units and re-tested for functionality.

▶ **Mixed integration** – sometimes referred to as sandwiched integration testing. This is a combination of the top-down and bottom-up testing methods. The top-level units are unit tested and then the bottom-level units are unit tested. They are then tested when they are integrated. This approach is particularly effective for very large products.

Performance testing

This is used to test the speed, response time, reliability, stability, scalability and resource usage of a software product/digital system under a specified workload. Sometimes referred to as 'perf testing', it is to ensure that any issues with performance are identified and rectified. So, in simple terms, it checks the speed, scalability and stability of the software program/digital system.

System testing

This is the testing of the complete and fully integrated software solution/digital system. This form of testing comes under **black box testing**. This means that the people testing the product do not have to have the knowledge associated with the inner design of the software, for example coding or development of the digital system. It is the final test carried out to verify that the software/digital system meets the required specification and tests the functional and non-functional requirements. The forms of system testing should be selected prior to the software being deployed to the stakeholders. There are various forms of system testing:

▶ **Usability/acceptance testing** – the testing of the application/digital system to confirm whether the end user will have a positive experience (or not). The software and/or digital system is tested to ensure it meets the requirements of the stakeholder(s). It also evaluates the compliance of the software and/or system with the requirements of the business and confirms that the required criteria for the end users has also been achieved.

▶ **Load/stress testing** – a form of non-functional testing which provides information on how the software/digital system conducts itself under specific loads. (This was also covered in performance testing.)

▶ **Regression testing** – to confirm that any changes or additions to the coding of the software or the digital system has not had any adverse impact on any of the existing features.

▶ **Functionality testing** – confirms that the software/digital system performs and functions according to the user specifications.

▶ **Migration testing** – when a program or system is tested to see that it can be moved to a different digital system (e.g. a new platform or server). Migration testing checks that data and any other dependencies can still be accessed, or if any conversion is needed.

▶ **Compatibility testing** – to test that the software will work on different platforms, environments and so on.

▶ **Boundary testing** – a form of black box testing (also referred to as specification-based testing). It is also used by developers conducting white box testing. This consists of a series of tests using the boundary values. An example is a program that is being developed to calculate examination grades. If it is deemed that, in order to achieve a distinction grade, a candidate must score no less than 75 marks and a maximum of 100 marks, then the boundary test would be to test that the calculations only produce a distinction grade within the 75 mark and 100 mark boundaries. So there would be a test for someone achieving 75 marks and someone achieving 100 marks. Any errors can be rectified and then re-tested.

▶ **Fuzz testing** – used to test the software by introducing invalid, random and/or unexpected data as inputs to software to see what happens.

Key term

Black box testing: testing of the software when the internal structure and design is not known to the tester.

Test yourself

1 Explain the term 'automated testing'.
2 Explain the purpose of concept testing.
3 Discuss the different forms of integration testing.
4 Identify the test that is used to confirm that the software will work on different platforms and in different digital environments.
5 Explain the term 'boundary testing'.

2.8.3 How automated and functional testing tools can be applied to test digital systems and code

In this section you will learn about automated and functional testing and what to consider when selecting tools. First, it is important to understand the terms.

Automated testing

This is where specialised tools are used to control the execution of tests. The actual results are compared with the expected results. Regression tests, which are repetitive actions, are automated. This is when functional and non-functional tests are re-run to ensure that any newly developed code has not introduced any bugs into the original software that had previously been functioning as intended. Testing tools are used to perform regression tests and automate data setup generation, graphical user interface (GUI) interaction, defect logging and product installation. Automation tools are used for functional and non-functional testing.

Functional testing

This is used to confirm the functionality of the software system against the functional requirements (specification). Each function of the software application is tested by providing an input and comparing the output against the functional requirements. Functional testing is primarily black box testing and does not deal with the application's source code. Functional testing checks the UI, application programming interfaces (APIs), database, security, client–server communication and so on. This form of testing can be carried out manually or by using automation.

When selecting tools for automated and functional testing several things need to be considered:

▶ **Compatibility with OSs** – the automation tool should support a variety of OS configurations so that, regardless of the OS being used by the end user, all possibilities have been tested and it has been confirmed that they can support the running of the software.
▶ **Versatility** – it is important that the tool selected supports the type of automated testing to be conducted. This should include functional testing, load testing and unit testing.
▶ **Compatibility with a variety of platforms** – it is important that the tool can support the range of applications and platforms that may be used by the end user. It is also advisable to consider any potential platform changes that may arise in the future.

▶ **Test creation** – a tool should provide more than one way to complete a task. This allows for tests to be conducted quickly and efficiently, regardless of the expertise and technical skills of the testers.
▶ **Maintenance** – maintenance can be simplified if the tool can generate modular test cases. These tests can easily be amended and used again. Detailed failure logs should also be available so that any issues can be easily identified and rectified.
▶ **Cost** – it is always important to consider whether a particular tool is value for money. Consider the price of potential upgrades, support fees and add-ons that may be required in the future.

Test yourself

1 Explain the term 'automated testing'.
2 Identify three considerations when selecting automated and functional testing tools.
3 Explain functional testing and what aspects of a digital system it is used to test.
4 Explain the term 'regression testing'.
5 You have been asked to select automated and functional testing tools to test some new software. The software will be deployed to a variety of global locations where the different offices are located. Discuss what considerations you need to take into account when selecting the testing tools.

2.8.4 How to apply root cause analysis to solve problems

The purpose of RCA Is to get to the 'root' of the problem. Asking questions such as:

▶ what is the problem
▶ what has caused the problem
▶ how can the problem be solved?

Once this is established, an assessment of the situation can take place and consideration given to what lessons have been learnt and how similar problems can be mitigated in the future.

There is a clear process that should be followed when conducting an RCA to ensure that the analysis is efficient, effective and useful. Whilst it is important to solve the problem, it is just as important to establish what is causing the problem and how this can be addressed. This is especially important when a business needs to continue to function. There is very little point in resolving a problem only for it to keep reoccurring and potentially disrupt operations.

A problem can be simple (one root cause) or it can be more complex (a number of root causes) and time must be given to establishing if all of the root causes associated with a problem have been identified. It is always important to focus on the problem itself with respect to how and why it happened as opposed to 'who did what'.

Problem solving requires a careful and methodical approach. It is more effective to plan the approach to investigating the problem and follow a step-by-step process. This enables you to gather relevant and detailed evidence as to why and how the problem has occurred.

Documenting RCA is particularly important as there has to be sufficient information to support the identification of any potential corrective actions that will be required. The implementation of corrective actions can also give an indication as to how future problems can reoccur. Decisions can be made as to whether there are adaptations or new processes, equipment, upgrades etc required as well as when they would be required to ensure that the problem does not arise in the future.

The Five 'Whys'

This method for performing RCA is known as the Five Whys approach (sometimes written as 5 Whys or 5 Ys). This is when for every answer to a 'Why' question, a further and deeper 'Why' question is posed. Eventually the root cause of the problem will be exposed. If there is more than one root cause, then the 'Whys' can look like a linked matrix. Consider this simple example:

Problem – I've got three points on my driving licence.

Why 1 – why have you got three points on your driving licence?

Response to Why 1 – because I was speeding.

Why 2 – why were you speeding?

Response to Why 2 – because I did not notice the speed restriction sign.

Why 3 – why didn't you notice the speed restriction sign?

Response to Why 3 – because I was not looking at the road signs.

Why 4 – why weren't you looking at the road signs?

Response to Why 4 – because I was not concentrating on the road signs.

Why 5 – why weren't you concentrating on the road signs?

Response to Why 5 – because I was too busy talking to my friend on my mobile phone using hands free.

So, the root cause of the problem is because the driver was too busy talking to their friend on their mobile phone when driving, even though they were using hands free. This meant they weren't concentrating on their driving, resulting in three penalty points on their licence.

The solution (and to prevent this from happening again) is not to use a mobile phone at all when driving, even if using hands free.

When to use RCA

RCA should take place when an issue occurs that results in outcomes that disadvantage an organisation and its stakeholders. Some criteria that can be used to determine if an RCA should be carried out are:

- ▶ failure of service delivery/functional operations
- ▶ loss of data
- ▶ the occurrence of an undefined process
- ▶ system downtime
- ▶ complaint or feedback from a stakeholder.

How to use the RCA process

To carry out root cause analysis, there is a clear process that should be followed:

- ▶ It is important to identify what the problem is and the impact that it has on the stakeholders involved. How does it impact on the function of the business? If whilst investigating the problem, there is no immediate indication of an impact on the stakeholders, it is always important to consider if there is the potential for an impact to happen.
- ▶ What information is there available about specific issues/problems that have occurred? When did it happen, what activities/tasks were being carried out at the time, how did it happen? It is important to gather as much data and information as possible. This can include system logs, feedback from all relevant stakeholders internal and external to the organisation as well as identifying where the problem has occurred before. Basically anything that can help you investigate the problem.
- ▶ What would be the impact on an organisation and its stakeholders if the problem was not resolved at all, or if there was a delay in resolving the problem.
- ▶ Once you have gathered the information and data you need, you can then investigate the problem and identify what caused the problem to occur. Remember there can be more than one cause to a complex problem.

▶ Remember to use the 5 Whys approach i.e. you ask a 'why' question and for the answer, you ask further 'why' questions until the root cause of the problem is exposed.

▶ If there is more than one cause to the problem, then a systematic approach should be taken. Do not try and address all the causes at the same time. Prepare a list of priorities with respect to the causes and slowly work your way through them. When prioritising the causes, consider the impact and resolutions that are required and always consider what effect this has on the functioning of the business. Some causes may be able to be addressed more quickly with very little, if any, downtime that will disrupt the running of the business. Others may take long and require e.g. a system to be taken down. Consider when it would be best to carry this out to minimise any disruption to the business processes. Safety can also play a big part in this and therefore there may be occasions when systems must be put on stop until the issue has been resolved.

▶ The focus is to eliminate the problem and therefore suitable solutions must be identified. The results of the investigation and the prioritisation of the tasks involved will assist in the identification of suitable solutions which can then be implemented. As previously stated, consideration must be given to timescales such as when each solution can be implemented and who will implement the solution.

▶ As with all situations when there are changes made to any form of digital technology whether it is hardware or software related, there is always a requirement to monitor the process and make changes as and when required.

▶ The final step in the process is to establish how and when the system will be tested/monitored to ensure that it is performing as required.

Test yourself

1 Describe the core principles of RCA.
2 Identify the three goals of RCA.
3 Explain the Five Whys.
4 Identify two situations when a RCA should be considered.
5 Discuss the steps that should be taken when carrying out a RCA.

2.8.5 How to construct an effective test plan

Testing a digital system or software should be conducted iteratively (during the construction process) and not just as a form of end-testing. It is important that the test plan is initiated with the planned tests at the start of the project. It does not mean that these are the only tests that are carried out, as the purpose of a test plan is that further tests can be added to it during the development of the digital system and/or software. Good test plans must be informative and allow for an analysis to take place on the overall performance of the digital system/software in line with stakeholder requirements.

An effective and informative test plan should do four things.

Identify the tests to be carried out

What should be tested depends on what exactly is being tested, for example a digital system and/or software. In addition, consideration should be given to the selection of the tests in line with the stage of the development of the system and/or software.

Describe the purpose of the identified test

It is important to ensure that the reason behind the test is clear and informative. If it is to test compatibility between the developed software and the OS and applications already in place, then each OS and application would need to be tested separately for compatibility. Do not group tests together; they must always be identified separately, and the purpose of the test clearly explained.

Identify test data to be used

There are many stages to the testing process, and it can be quite complex. One of the key activities is the creation of test data in preparation for the tests to be carried out. Each type of software testing requires data that is different, sufficient and relevant. The amount of data required for testing performance, stress and load testing of digital systems is very large.

The types of data are categorised as:

▶ **Valid** – data that would normally be used while the software is being used by the end user. It confirms that the processing of the data is compliant with the requirements and that it is processed and stored as intended.

- **Valid extreme** – in relation to any parameters that have been set for the software, for example an input field may be set to take values between 5 and 50 only. So a test is carried out to confirm that if a value of 50 is entered, and if a value of 5 is entered, that the system processes and stores the data as intended. It can then be assumed that all other values between 5 and 50 will also be processed and stored as intended.
- **Invalid** – to test how the software processes the input of any invalid values, for example showing messages to notify the end user that the data that has been input is not appropriate.
- **Invalid extreme:** this is data that falls outside of the boundary of extreme values. E.g. If the extreme values were 5 – 50 and the data was <=4 and/or >=51, then this is classed as invalid extreme.
- **Erroneous data:** this is data that should not be accepted by the system and therefore should not process it and will create an error message. This data can also include data input in the incorrect

format e.g. a date require in DD-MM-YYYY that has been input as YYYY-MM-DD.
- **Absent** – it is always important to test how the software reacts if there is data missing. If the missing data prevents the software from carrying out its processing and storage function, then an error message should be seen.

Describe the expected results

The expected result is the ideal result that should be obtained after the test has been carried out. It is important that the expected result is described in detail and it may include more than one possible result. The person conducting the testing must be able to compare the expected result with the actual result of the test and establish whether there is an issue that requires rectifying. Therefore, it is important that tests are not grouped together as the expected results must be clearly defined.

Depending on the organisation, a test plan can be an extremely detailed formal document or a simple plan documenting the tests as in Table 2.14.

Test no	Description of test	Type of test data	Expected results	Actual results	Resolution	Re-test no

▲ **Table 2.14** A simple test plan

Test yourself

1 Explain the term 'erroneous data'.
2 Explain why it is important that tests are not grouped together.
3 Discuss the different types of test data and how they are used.

4 Large quantities of test data are required when testing the performance, stress and load capability of a digital system. True or false?
5 On the test plan given in Table 2.14, there is a column for the 'Re-test no'. Explain the importance of re-testing after rectifying an error.

Skills practice

An online retailer applies delivery costs to each order. The delivery cost is calculated on the total cost of the items bought. The minimum order is £10. The table shows how the delivery costs are calculated.

Cost of items	Delivery cost
£10.00 – £24.99	£2.50
£25.00 – £39.99	£5.00
£40.00 – £59.99	£7.50
£60.00 – £74.99	£10.00
£75.00 or above	Free

The retailer stores the customer delivery addresses, the items they have purchased, the total cost of the items and delivery costs.

As part of the Skills practice at the end of Content area 1, a top-down diagram and algorithms were created and tested.

The retailer has analysed the top-down diagram and the algorithms created and wants to move the project to the coding stage.

The retailer has asked that customer details be retrieved from, and saved to, a text file. When the customer placing the order is an existing customer their details will be retrieved from this text file. Updates to existing customer details and new customer details will be saved to the text file.

You have been asked to:
► Create the code based on the top-down diagram and algorithms created in Content area 1.
► Use input validation to reduce the possibility of errors.
► Ensure the code conforms to PEP8.
► Select the testing and quality assurance methodologies to be used to test the code, including a justification for your choices.
► Create a test plan.
► Use the test plan to carry out testing, correcting any errors found, and re-testing where necessary.

Assessment practice

1 Identify the most appropriate data type to store a mobile phone number. Justify your choice.
2 Describe how the array data structure can be used in code.
3 Explain, using an example, the difference between a local and a global variable.
4 Explain why variables should be given meaningful names.
5 How can text files be used as input to code?
6 Identify and describe one benefit and one drawback of using a bubble sort to sort a list of numbers into a defined order.
7 Identify the most appropriate validation technique for checking a UK postcode. Justify your choice.
8 What is the difference between backward and forward compatibility testing?
9 Identify and describe two different types of integration testing.
10 Describe, using an example, how boundary testing can be used during testing.

Content area 3: Emerging issues and impact of digital

Digital transformation is, as you know, achieved through technology. But you must also remember that the human aspect is as important as the technology. Digital professionals must understand the ethical and moral issues within the digital sector for a variety of business contexts. It is important to understand how technological developments impact on individuals, organisations and society on a global scale.

Digital technology is constantly evolving, and it is important that digital professionals keep up to date with any important and innovative developments such as the Internet of Things (IoT), Artificial Intelligence (AI), Augmented Reality (AR) and Virtual Reality (VR).

Learning outcomes

In this content area you will learn about:
3.1   Moral and ethical issues
3.2   Emerging trends and technologies

3.1 Moral and ethical issues

3.1.1 Ethical and moral issues, and organisations' and individuals' response to these challenges

There are many challenges faced by organisations and individuals as the reliance on technology increases. While it is important that all organisations that intend to implement digital transformation embrace **ethics**, it is more difficult on an individual level. This is because it is individuals within an organisation that make decisions, not the organisation itself.

What is seen to be ethical can vary across:
▶ individuals
▶ groups
▶ religions
▶ cultures.

With a digital society that has expanded globally and is extremely fast moving, there is scope for a wide range of interpretation. Therefore, it is important that digital professionals at all levels determine:
▶ what is ethically correct (the right thing to do)
▶ the ethical training needs that may be required
▶ the ethical awareness within their organisation.

It is also important to consider the current legislation within the UK and other countries, and the impact and challenges these present to an organisation when it increases its reliance on technology.

> **Key terms**
>
> **Ethics:** the **moral** principles, or rules, that govern a person's attitudes and behaviour.
>
> **Morals:** the principles of what people believe is right or wrong.
>
> **Distributed denial of service (DDoS):** when a network is flooded with so much traffic that it cannot operate or communicate as required.

You are going to explore the challenges faced by organisations and individuals when there is an increased reliance on technology.

Acceptable use

Acceptable use is how organisations and individuals conduct themselves in a responsible, ethical and legal manner. Organisations develop an Acceptable Use

Policy (AUP) to promote the safe and responsible use of their digital technology, network and internet access while at work. Internet Service Providers (ISPs) require individuals to agree to an AUP. This usually requires individuals to agree not to:
▶ use the service for any illegal activities
▶ hack into the security of any computer network or individual system
▶ send spam or junk email to anyone who does not wish to receive it
▶ implement a **distributed denial of service (DDoS)** attack on a system by flooding the servers.

> **Activity**
>
> An organisation has asked you to provide them with information on the purpose of an AUP and what it contains. Research what would be included in an AUP. Create a guidance document for the organisation which clearly explains the role of an AUP. Include guidance on what an AUP contains with a justification for their inclusion.

Autonomous operation

Across a range of business sectors, technology carries out tasks and processes on behalf of people. This technology still needs the regular intervention of a person or people – for example, to tell the system what to do, to rectify faults or to make improvements.

However, systems can also be designed to operate autonomously. This means that the system will carry out the necessary tasks or processes without the need for, or with very limited, human intervention. Such a system can make its own decisions based on the data it collects – for example, to resolve, or even pre-empt, problems – and find ways to improve itself – by making processes more efficient and cost-effective.

The are several positive potential impacts of autonomous operation in businesses. For example, the customer experience can be improved by the business implementing chatbots. These are autonomous operations that enable customers to source answers to their questions and, in some instances, receive refunds or return faulty products. Chatbots constantly learn through their interaction with customers. This can improve customer satisfaction and customer loyalty. Stock control can also be conducted through autonomous operations. As the system is analysing information relating to stock in and stock out, it can automatically re-order items when they reach a pre-defined re-order level. In these examples, autonomous operations can help a business to operate more effectively and efficiently.

There are also several risks to consider when designing or implementing autonomous systems. Firstly, such systems can be biased. This might be caused by erroneous data being fed into the machine learning process. A system that is learning through biased data can cause problems that can seriously impact the operational functionality of a business, resulting in loss of revenue, loss of custom and, in some cases, serious risk to life. As with all things with digital technology, there is always the potential for 'Garbage In, Garbage Out'.

This can be reduced (and even eradicated) by ensuring that a wide range of tests are carried out and the results are carefully analysed. It is therefore important that when considering autonomous operations and machine learning the correct machine learning model is selected, the data used for the learning is carefully monitored and a wide range of tests are conducted prior to and during implementation.

Autonomous systems can also have a significant impact on certain sectors within the workforce as jobs may be 'taken over' by technology. Take, for example, the car industry. You will now find the use of robotics on the production line to carry out tasks such as welding, assembly and painting, all of which were previously performed by humans. It is therefore important that businesses prioritise the retraining of these employees so that they can be deployed in other areas of the business.

There are also ethical considerations. While autonomous systems can carry out tasks that may be dangerous to humans, they can also cause other, new safety issues. Consider the use of an autonomous system being used to control high-pressure die-casting of aluminium alloys. The process requires the application of pressure up to 160 MPa on the aluminium. The resulting molten metal is cooled until it freezes. There are serious consequences that can result in fires or even an explosion if the high pressure exceeds 160 MPa. Autonomous systems rely on data to learn the process and to carry out the task. Like many things where data is concerned, it is open to cyber-attacks. A hacker could access the systems and delete or amend the data and cause serious consequences, for example to push the pressure in high-pressure die-casting above 160 MPa.

It is therefore important that businesses ensure that human intervention can take place to prevent any harm to human life and/or the environment. Prior to implementation, there should be an assessment as to whether the technology is ethical, safe and, of course, legal. In addition, it must be understood by the people who will be working alongside it. Once the autonomous system is deployed, there should be a means for a human to take over control and over-ride any decisions which could cause harm to humans, the environment or, in extreme circumstances, society as a whole.

Changes in societal norms and the behaviours of individuals

These are the changes in people's beliefs, attitudes and behaviours as part of a social group, and as individuals, due to the increasing reliance on digital technology. Making positive changes is a challenge. Many people are reluctant to change because of a lack of knowledge and in some cases a fear that their jobs may be at risk.

Digital technology has transformed the way that people interact with each other and see the world. For example, the Internet, smart devices and recent advancements in networking have changed the way that we communicate with each other, especially through social media.

There are many positive sides to this transformation. The use of Voice over Internet Protocol (VoIP) and social media means that people can maintain contact with friends and family more easily. Organisations also use the advances in digital technology to interview potential employees or clients on a global scale.

Learning behaviours have also changed. People can 'meet' people from other countries and learn about their culture. Learning a language can now be accessed online instead of attending a classroom.

However, whilst digital technology has assisted in bridging the global gap, it has also created issues. People, for example, might spend lots of time communicating with others who are in a different location, and yet not communicate effectively with people within their own environment.

> **Key term**
>
> *Societal norms:* unwritten rules about beliefs, attitudes and behaviours that are considered acceptable in a specific social group or culture.

Some of the positive and negative effects on society of digital technology are:

▶ **Positive effects:**
  – bridging the global gap
  – the internet provides access to a wide range of resources
  – faster and easier communication.
▶ **Negative effects:**
  – virtual distancing where people are physically together but detached from each other because they are immersed in the technology within their environment
  – reduced physical interactions through fewer face-to-face meetings and conversations.

Changes in the culture within an organisation

Organisational culture is the beliefs, assumptions, values and ways of interacting that contribute to the social and psychological environment of an organisation. As technology increases, changes to organisational culture must take place.

When an organisation's culture relies on digital technology and is fast moving, employees may occasionally use their own tools and solutions without contemplating collaborating and sharing information with others. This can create inconsistency where optimisation is prioritised over innovation. This reduces the culture of experimentation and can prohibit the growth of the organisation.

Environmental issues

The increased use of technology within businesses and within homes can have both negative and positive impacts on the environment. It is therefore important to consider whether the positive impacts outweigh the negative ones. For example, while digital technology has become more efficient, by supporting the development of more efficient and effective processes, it is still important that consideration is given to reduce the associated energy consumption.

An example of the positive impact of the use of technology on the environment is in relation to remote working. Employees can access the information they need remotely, attend online meetings and collaborate on projects. There is therefore a reduction in greenhouse gases because they do not have to use their cars to make the daily journey to the office. Businesses that have offices in other countries can hold online business meetings without the need for the participants to travel between countries to attend.

A negative impact is that as technology has become more accessible and affordable for businesses and society to use, the consumption of power has radically increased because more and more people and businesses are using technology. Many systems are running 24/7, whether this is in-house IT systems, the systems provided by cloud providers or someone using their mobile phone to chat online. Moreover, many people are investing in using technology to turn lights on and off through voice control, and many televisions and home computers are left on standby. All of these have increased the consumption of power. This energy has been primarily generated using fossil fuels, which are known to create vast amounts of pollution and contribute to global warming.

What is more, the manufacturing of components used in technological equipment includes the use of non-renewable sources such as gold and other precious metals. If there is the constant use of non-renewable resources to manufacture the components required for the technology, then there is a danger that they will eventually run out altogether. There will only be a finite amount of supply and eventually with constant use it will not be available. A non-renewable resource cannot be replenished.

In addition, technological equipment tends to contain non-biodegradable materials such as plastic and toxic material such as cadmium, lead and mercury. The **e-waste** that is generated from discarded technological equipment therefore creates an environmental problem. If this e-waste is not dealt with appropriately and just thrown onto landfill sites, the toxic materials will eventually leak into the ground. This can contaminate the surrounding habitat such as water and plants that, in turn, can poison people and animals that inadvertently consume them.

Key term

e-waste: this is an abbreviation for electronic and electrical waste. This relates to all forms of electrical and electronic equipment (EEE), and the components they are made of, that has been discarded by the owner with no intention of re-use or re-purposing.

Research

In the early 2000s, the development of wireless technology and the increased use of smartphones brought about a major change in how people controll their finances.

Research the use of technology within the banking sector in the last 10 years and consider:
▶ the positive and negative impacts on the industry in general
▶ the positive and negative environmental impacts of the increased use of technology within the industry
▶ the ethical and moral considerations of the use of technology within the industry.

Globalisation

The spread of technological globalisation is the spread of technology across borders. This includes the borders of different countries and continents. This, in itself, creates challenges for different countries. Is the digital technology readily available; does the country's infrastructure support an appropriate level of internet and Wi-Fi access? This has links with what is termed the digital divide.

Further information on the digital divide can be found in Content area 4, section 4.2.3.

Inclusion and diversity

It is important to accept that everyone is different. People have different skills and differing needs. This creates the challenge of how to include them in the workforce, how to make them feel valued and how to provide equal access to all.

Refer to the Equality Act in Content area 4, section 4.1.4.

Diversity is the range of people in the workforce. This can mean people of different ages, religions, ethnicities, genders, sexual orientations, and those with disabilities. Diversity also means valuing those differences.

An inclusive workplace means that everyone feels valued and feels safe to:
▶ put forward their ideas
▶ raise issues and suggestions
▶ try to do things in a different way (with the approval of management).

Organisations should develop an equality, diversity and inclusion policy which takes into consideration the increased reliance on digital technology. The aim of the policy is to:
▶ provide equality, fairness and respect for all employees, whether temporary, part-time or full-time
▶ not unlawfully discriminate because of the Equality Act 2010 protected characteristics of age, disability, gender reassignment, marriage and civil partnership, pregnancy and maternity, race (including race, nationality, and ethnic or national origin and religion) and sexual orientation
▶ oppose and avoid all forms of unlawful discrimination. This includes in respect of pay and benefits, terms and conditions of employment, dealing with grievances and discipline, dismissal, redundancy, leave for parents, requests for flexible working, and selection for employment, promotion, training or other developmental opportunities.

Refer to the Equality Act in Content area 4, section 4.1.4.

Research

Search the internet for a sample equality, diversity and inclusion policy. (Hint: ACAS provide a sample template.) Create a policy for an organisation that you know or have researched. Make sure that you are considering the increase on the reliance of digital technology within the organisation.

Monitoring of employees

Employers may want to monitor their workforce for various reasons, and legislation does not prevent them from doing so. But employers must remember their workforce are entitled to some privacy at work and inform them about any monitoring arrangement and the reason for it.

Refer to monitoring systems in Content area 4, section 4.1.6.

Open source and Creative Commons

Open-source software is software where the source code is made available for any person to inspect, modify or enhance, subject to having an open-source licence. Source code can be amended by programmers to change how a piece of software works. They can improve it by adding additional features or rectify any errors which prevent it from working properly.

Any type of creative work can be made available in the **public domain**. For creative works, that is writing and multimedia, copyright owners can apply one of the **Creative Commons (CC)** licences. CC licences are not the same as copyright. Copyright gives creators exclusive rights over their creative works, for example the right to reproduce, display and make adaptations. The phrase 'All Rights Reserved' is used by creators to emphasise that they reserve all rights granted to them under law. CC licences give creators several choices ranging from retaining all rights to relinquishing all rights. This is often referred to as 'Some Rights Reserved'.

> ### Key terms
>
> **Public domain:** belonging to or being available to the public as a whole.
>
> **Creative Commons (CC):** international, not-for-profit organisation that provides free licences for creators to use when making their work publicly available. The licences provide permission for others to use the work under certain conditions.

The collection and use of data

Legislation such as the Data Protection Act within the UK governs what organisations can do with respect to the collection and use of data. Large international organisations have to comply with the legislation for each country where they conduct their business.

> Further information on legislation can be found in Content area 4.

Unequal access to technology and/or digital services

Known as the 'digital divide', this is a global issue. It relates to the difference between those who have access to technology and those who do not. The divide can be caused by lack of money, internet access, IT literacy and location.

> Further information on the digital divide can be found in Content area 4, section 4.2.3.

> ### Test yourself
>
> 1 Identify one positive and one negative effect digital technology has on human behaviours in the workplace.
> 2 Explain the term 'open source'.
> 3 What is the purpose of Creative Common licences?
> 4 What is another term used for unequal access to technology and/or digital services?
> 5 Describe two things that an organisation must do if they want to monitor their employees.

3.1.2 How organisations and individuals respond to ethical and moral issues when designing and developing digital systems

Use of guidelines from professional organisations

It is important to understand the role of professional organisations. Some of their functions include:

▶ issuing a code of conduct to guide professional behaviour
▶ promoting fairer access to the profession to people from all backgrounds
▶ dealing with complaints against professionals and implementing disciplinary procedures
▶ setting and assessing professional examinations
▶ providing support for continuing professional development (CPD)
▶ publishing professional journals or magazines
▶ providing networks for professionals to meet and discuss their field of expertise
▶ representing their members in lobbying government about relevant legislation
▶ providing careers support and opportunities for students, graduates and people already working in the profession.

In Content area 4 you will find further information about the codes of conduct and guidelines produced by professional bodies. Codes of conduct are a collection of written rules, principles and values, as well as employee expectations, behaviours and relationships, that organisations believe are important and significant to ensure the successful operation of the business.

A written code of conduct provides guidance for employees, customers and other relevant stakeholders. It confirms what is most important, valued and desirable with respect to relationships, interactions and the organisation's global perspective.

The code of conduct is sometimes referred to as a 'Code of Business Ethics', 'Code of Business Ethical Conduct' or 'Code of Ethics and Standards'. Whatever its title, it serves as a framework for ethical decision making within an organisation. It is used as a communication tool informing internal and external stakeholders about the values of an organisation. It informs employees, customers, suppliers and partners about how they can expect to be treated.

> Refer to the section about the Equality Act in Content area 4, section 4.1.4.

Strategic planning and decisions

When setting an organisational strategy for the design and development of digital systems, it is important to include a strong sense of ethics, as well as considering the well-being of the current stakeholders and society at large. This is an integral part of the strategic planning process and is often referred to as Stakeholder Theory. Organisations usually appoint a board of directors who oversee the strategy to ensure that it remains aligned with the organisation's ethics and values.

When strategically planning and making decisions for digitally transforming an organisation, it is important that it is a long-term plan which focuses on:
- the implementation of new technologies
- integrated digital media channels
- smart ways of working in a digital environment using the available technology.

A strategic plan documents the mission, vision and values of an organisation, as well as its long-term goals and how they are to be achieved. A carefully thought-out strategic plan can play a key role in supporting the growth of the business and its success. This is because it informs the organisation and its employees how best to respond to opportunities and challenges.

It is important to include ethics in strategic planning to ensure that all areas of the organisation support its overall ethos and values. Therefore, when considering the organisation's ethics, all **stakeholders** should be taken into account, and the key ethical considerations included in any strategic plan:
- stakeholder participation
- organisational values
- individual values
- managing change.

Managing change due to a change in strategy has an impact on organisational structure. This change can be managed in different ways. It is important

that any change is managed in a way that supports organisational and personal values. It is also important to consider:
- Who will gain from the change?
- Who will not gain from the change?
- Is the change worth the risk?
- How will the organisation work with those affected by the change in order to help them make the transition?
- How will the organisation work with those who cannot make the transition?

Key term

Stakeholder: anyone with an interest in a business or organisation. Stakeholders can be individuals, groups or other organisations, or businesses that are affected by the organisation's activity. They can include customers, suppliers, employees, communities, government and even the ecosystem.

The content of internal policy documents

Internal policies communicate the organisation's values and its expectations with respect to the behaviours and the performance of its employees. They are often called workplace policies and define what is acceptable and unacceptable behaviour in the workplace. They also include the implications if the policies are not adhered to.

Examples of internal policies include:
- annual leave and sickness
- IT security and data protection
- maternity, paternity and parental leave
- smoking
- dress code
- whistleblowing
- bullying and harassment
- health and safety
- social media
- acceptable use
- diversity and inclusion.

Research

Research the different internal policy documents used in your workplace. Prepare a report, explaining the purpose of each document and the acceptable and unacceptable behaviour required from employees.

Good policies are important to organisations as they outline the intentions, values and long-term goals for the employees.

Company culture and how this is established, communicated and sustained

A company's culture is the shared values, beliefs and expectations that assist in guiding, motivating and informing the actions of its workforce. Successful companies tend to be ones that have a strong company culture. This helps the workforce to interpret and understand the company and its goals, which in turn encourages active participation, a sense of community, teamwork and positive performance. This can improve the way that people interact with each other and the mutual respect that is demonstrated across the business, as well as helping employees to carry out their tasks efficiently and effectively.

So, how can a company establish a positive workplace culture? As with all things, it starts at the top: with the behaviours of the management team. These are often referred to as **leadership behaviours**. Leadership behaviours include the articulation of clear and compelling visions for the company that relates to all employees, as well as guiding and inspiring the workforce to engage with this vision through their roles and to meet the strategic goals of the company.

> ### Key term
>
> **Leadership behaviours:** the mannerisms and actions that make people effective leaders.

Good leadership behaviour means leading by example. This will motivate employees and lead to good staff relationships that have a positive impact on the culture of the company and on the company as a whole. Positive employee relationships help to motivate the workforce and encourage them to work hard on their tasks. It also helps to retain employees as a happy workforce has no need to look for work elsewhere. Managers should lead by example and if there is a requirement for certain teams within the company to work extra hours to complete an urgent job, then the managers for those teams should also be seen to be working late alongside their team to help get the job done.

Communication within the workplace should not be one-sided. The workforce should be given the opportunity to:

- put their ideas/suggestions forward for consideration in relation to working processes and practices. People

who constantly carry out the same tasks on a daily basis can often identify changes that can enhance production by saving time and money
- give feedback on the current company culture, highlighting the positives and the negatives. By analysing the feedback it is possible to identify areas within the company culture that need addressing and consider how these changes can be implemented for the positive benefits of everyone.

By encouraging communication between management and the workforce, a positive company culture can be sustained.

It is also important that companies consider the health and well-being of their workforce; people need to be healthy to work and to enjoy their lives outside of work. Companies should provide the resources to help the workforce protect and maintain their health, both in terms of their physical safety/comfort and their mental/emotional well-being. There are rules and regulations in place that must be followed for example, when people use digital equipment. (See Content area 4.)

The design and implementation of new digital systems invariably has an impact on the workforce. People may become fearful for the security of their jobs and whether they are going to be made redundant. They may also be concerned with having to acquire new skills to use the new technology. To maintain a positive company culture, it is important that these potential changes are communicated effectively to the workforce in advance of any changes. There should be clear explanations of the reasons for the changes, the impact they will have on the company and the workforce, as well as how the workforce will be supported to embrace them. Companies should also take the opportunity at an early stage to seek feedback from the workforce on the potential changes, asking them for ideas, suggestions and even their concerns. Management should analyse any feedback from the workforce and communicate any reassurance that may be required or provide further explanations. Where appropriate, management should also act upon employees' feedback; a positive company culture is one in which feedback is considered and acted upon, as opposed to asked for and ignored.

The key factors in fostering a positive company culture are:

- clear, inclusive and engaging explanations of the company's culture and its goals
- mutual respect between the management team and the rest of the workforce
- continuous communication between management and the workforce

▶ transparency and prompt communication of potential changes

▶ development of workforce participation by encouraging them to put forward ideas, suggestions and even concerns

▶ provision of resources that will support the health and well-being of the workforce.

Whistleblowing

Many organisations have whistleblowing policies. A whistleblower is a worker who reports certain types of wrongdoing. The wrongdoing must be in the public interest. This means that it must have an impact on other people. A whistleblower is protected by law and must not be treated unfairly or lose their job due to reporting a wrongdoing.

Types of workers are:

▶ employees

▶ trainees

▶ agency workers

▶ a member of a limited liability partnership (LLP).

Whistleblowing complaints include:

▶ a criminal offence such as fraud

▶ danger to the health and safety of a person or group of people

▶ risk of or physical damage to the environment

▶ the organisation is breaking the law, for example does not have appropriate insurances in place such a public liability insurance

▶ someone is covering up (hiding) a wrongdoing

▶ there has been a miscarriage of justice.

Whistleblowing complaints do not include personal grievances unless it is in the public interest. Personal grievances include:

▶ bullying

▶ discrimination

▶ harassment.

These would be reported under the employer's grievance policy.

> ### Key terms
>
> **Agency worker:** a person who has a contract with an agency but works temporarily for someone who hires them.
>
> **Limited liability partnership (LLP):** a company owned by two or more people. Each person pays tax on their share of the profits but is not personally liable for any debts the company cannot pay.

> ### Test yourself
>
> 1 Identify four ethical considerations when developing a strategic plan.
> 2 Explain the term 'code of conduct'.
> 3 Identify four whistleblowing complaints.
> 4 Describe the positive results if an organisation has a strong, positive culture.
> 5 Explain how employers can create positivity in the workplace.

3.1.3 How individuals use a range of observational techniques to inform situational awareness

Situational awareness

Situational awareness is simply knowing what is going on around you. Situational awareness is important to everyone as we should all be aware of our surroundings and the potential hazards we face. It is the responsibility of each of us to look out for the health and safety of ourselves as well as the people around us.

There are three levels of situational awareness:

▶ Level 1 – to hear, see, become aware of something using observational techniques

▶ Level 2 – understanding what you have become aware of

▶ Level 3 – using what you understood to inform future actions.

Observational techniques are:

▶ **Observing normal behaviour** – this is being able to establish a person's usual behaviour within the working environment. By knowing a person's normal behaviour, you can become aware when they start to behave out of character. You should also be aware of your own normal behaviour as this can help you to identify if you also start behaving differently. This is referred to as self-awareness.

▶ **Awareness of co-workers** – it is important to be aware of your co-workers and be able to notice if something appears wrong. Are they stressed, are they ill, are they struggling to carry out their work? These are just a few of the things you may notice with a co-worker. Once you have established the issue you can take appropriate action, for example offer to help them with their work, get medical help and so on.

▶ **Recognising change or abnormal behaviour** – through being able to understand a person's normal behaviour you can easily become aware if there is a change. What future action you take will depend on what you interpret as the cause.

> **Test yourself**
>
> 1 Explain the term 'situational awareness'.
> 2 Identify the three levels of situational awareness.
> 3 Describe the term 'observational techniques'.
> 4 Explain the term 'awareness of co-workers'.
> 5 Describe the term 'recognising changing or abnormal behaviour'.

3.2 Emerging trends and technologies

3.2.1 How developments in digital technologies impact on organisations, individuals and society

The increase in the use of digital technology has changed the way that people and society connect and collaborate in the workplace and/or on a personal level.

There are positive impacts with the increased levels of communication and social interaction in relation to time, social context and boundaries. We are now able to communicate with people from other countries and learn about their culture and countries, therefore expanding our own knowledge and understanding. New ways of working and learning (education) have also become more accessible due to digital technology.

However, the increase in digital technology has also had negative impacts. Research has shown that the excessive use of digital technology by humans has had a negative impact on **cognitive** and behavioural development and in some instances mental and physical well-being. There are changes to social interaction, as any face-to-face communication and collaboration is often substituted by online interaction. Increased digital technology in the workplace can result in the fragmentation of jobs, threatening job security.

It is therefore important that the opportunities and risks associated with the increase in digital technology are understood. Individuals, society and organisations need to learn how to exploit the benefits of digital technology while working to reduce any negative impacts.

> **Key term**
>
> **Cognitive development:** how we think, explore and work things out. It is the development of our knowledge, skills, understanding and ability to solve problems.

The Internet of Things

The IoT is a system of interrelated, internet-connected objects that collect and transfer data over a network without human intervention. It is built on a connection between people, processes, data and things. These are called the pillars. Each pillar increases the capabilities of the other three pillars. Let's look at each of the pillars.

People

The way people connect to the internet has changed dramatically over the last two to three decades. People now use wearable technologies and smart devices which have transformed the way we obtain and share information. There are self-monitoring devices such as the 'Fitbit' that allows people to track exercise, monitor the function of vital organs such as the heart, and even the quality of their sleep. It is the connecting of people in the most relevant and valuable ways.

Process

The Internet of Things (IoT) is a network of physical objects. These objects are the 'things'. They are embedded within software, sensors and other technologies, which enables them to connect and exchange data with other devices and systems over the internet.

The IoT has revolutionised the way organisations work and how they manage their business. Think of an e-commerce company such as Amazon and how people's shopping habits have changed by buying online, combining the convenience of shopping from home with fast delivery times. (Of course, this also has a negative impact on the high street stores.) The process comprises the delivery of the correct information to the correct person/machine at the correct time.

IoT sensors are used in warehouses by businesses like Amazon. They use IoT sensors to monitor the operating conditions of equipment and machinery and the data captured is processed and analysed by advanced machine learning algorithms. This enables the operators to predict failures of equipment e.g.

forklift failure and other critical handling equipment. Warehouse managers can then address the problems promptly, minimising the risk to a reduction in productivity by keeping costs and downtime to a minimum.

Data

This is about maximising data into more useful information for decision making. The world is flooded with data and one of the main functions of the IoT is to gather huge amounts of data to improve operations and functionality. Data could be hours of footage from a surveillance camera, how much exercise you have done in a day or what TV programmes you watch. These are examples of individual data but imagine the vast amounts of data that can be collected, processed and analysed from organisations on a global scale. All you have to do is consider social media platforms such as Facebook, Twitter, WhatsApp, LinkedIn, Instagram, TikTok and so on.

Digital marketing benefits from the IoT by helping to automate the processes involved with social media marketing. It can generate automated posts and gather and analyse data that would be difficult to gather as quickly by standard means such as questionnaires etc. Due to the global reach of social media, the accessible customer base is therefore much greater and the potential for comprehensive feedback also increases. Personalised advertisements are already becoming more common on social media and websites and this is due to the IoT gathering vast amounts of data for analysis.

Things

These are the physical devices and objects connected to the internet and each other for intelligent decision making. There are many devices that we can interact with on a daily basis that are connected to the internet. We use our mobile phones for online banking, paying for goods or just checking the weather. We can control the lighting and heating in our homes when we are away.

The Internet of Everything

The IoT is the connection of physical objects (things). These 'things' can be sensors, devices etc and are often defined as a type of Machine to Machine (M2M) communication technology. The Internet of Everything (IoE), however, comprises of people, data, things and processes. These are commonly referred to as the four pillars of the IoE. IoE is an extended version of IoT. People use things. These things collect and produce

data that is processed. The data that is processed allows people to make smarter decisions that are data driven.

The personal and business possibilities from the IoT are vast. Businesses are motivated by the use of IoT due to the opportunities for increasing profits, reducing operating costs and improving the efficiency of the business. The IoT provides data which can be analysed to:

▶ automate work processes and streamline workflow
▶ develop visualisation patterns of usage
▶ enable organisations to effectively compete in a continuous changing business environment.

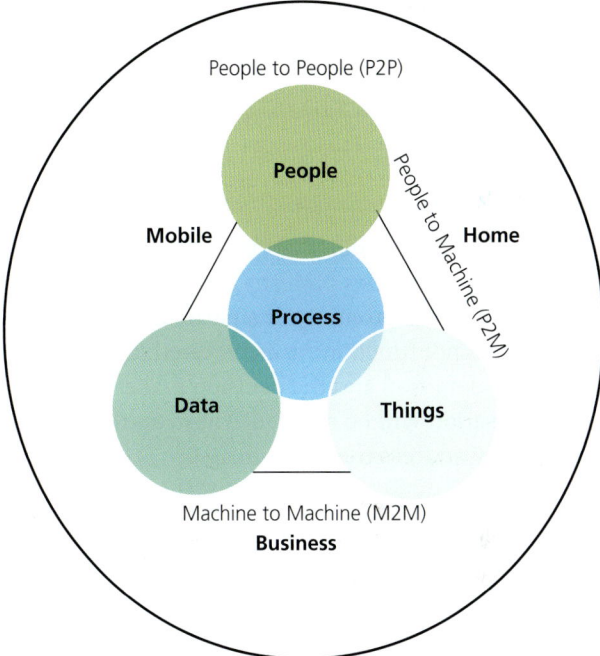

▲ Figure 3.3 The internet of everything

Research

Research two different business sectors, for example retail and entertainment, and how the IoT has been implemented. Produce an infographic highlighting the key outcomes from your research.

AI, machine learning and deep learning

First of all, you need to understand the difference between AI, machine learning (ML) and deep learning.

▶ **AI** is the development of computer systems to perform tasks that would normally require human intelligence. This can include speech recognition, visualisation or visual perception, decision making and translations between languages. Therefore, it is the development of computer systems to think and work like humans.

▶ **Machine learning** is an application of AI where computer systems automatically learn and improve with experience without being programmed. The focus is on the development of computer programs that facilitate access to data that is used to allow a computer system to learn for itself.

▶ **Deep learning** is a function of AI that imitates how the human brain works with respect to the processing of data and creation of patterns in order to make decisions. This is sometimes referred to as deep neural learning or deep neural network.

Because ML can access and learn from vast amounts of data, it is used effectively in the manufacturing sector. Machinery is linked to the network and there is a constant stream of data being provided with respect to the functionality and production rate of each machine. This is sent to a local point for analysis. Because this data can be so vast it is difficult for a human to analyse it all quickly and therefore critical situations can be missed. ML quickly analyses data, identifies patterns and can therefore identify anomalies promptly. If a particular piece of machinery is not functioning as required, this will be quickly identified and the decision makers within the organisation will be notified. They can then immediately address the problem.

Deep learning is a more specific version of machine learning and uses neural networks to facilitate **non-linear thinking**. This is critical to performing more advanced functions by analysing a wide range of factors simultaneously. Consider self-driving cars. Deep learning is used to contextualise information picked up from the car's sensors, for example the distance from objects, the speed of travel and a prediction of where they will be over time. All of these are calculated at the same time so that the car can make decisions rapidly, such as stopping quickly and changing lanes.

ML algorithms can tend to reach a state where there is little or no change. Deep learning models, however, continue to improve their performance as more data is received. Therefore, deep learning models are more scalable and detailed and far more independent.

> ### Key term
>
> **Non-linear thinking:** the ability to make connections and draw conclusions from unrelated concepts or ideas.

The challenges of AI for society and individuals

AI will require the workforce to evolve. Many people are fearful that jobs will be lost to machines. The challenge is to encourage humans to motivate themselves by taking on new responsibilities that require the unique abilities of humans.

The impact of AI on society will have economic, legal and regulatory implications that need to be prepared for. Consider, again, the self-driving car. Who is at fault if a person is injured coming into contact with a self-driving car?

It is important that AI does not become so good at doing the work intended that it crosses the ethical and legal boundaries. Any AI algorithm must be developed to align with the goals of humans.

Of course, there is also the issue of privacy being compromised due to the collection of data about every minute of every day of people's lives. If businesses and governments make decisions based on the intelligence they gather about people, it could devolve into social oppression.

Positive impacts of AI on society and individuals

AI can improve dramatically the efficiency of the workplace and the work humans do. AI can take over repetitive and dangerous tasks which allows the human workforce to carry out work that involves creativity and empathy. If a person carries out work that is more engaging for them they are happier and have job satisfaction.

AI has had a real positive effect on the healthcare sector as it helps to improve monitoring and diagnostic capabilities when dealing with patients. These improvements with respect to healthcare facilities can reduce operating costs and therefore save money which is needed to employ more staff. Personalised treatment plans for patients and drug protocols, as well as wider access to information across medical facilities which help to inform patient care, are important positive aspects of the implementation of AI.

AI in organisations

AI can save businesses time and money, as well as increasing productivity and improving operational efficiency. This is achieved by automating and improving tasks, workflow and processes.

AI includes the use of cognitive technologies that can make faster business decisions based on resulting outputs from other processes. Cognitive technology is a product within the field of AI that is trained to think and behave like a human. Think of the use of robotics for building cars. They can carry out the actions a human would do and do not need to stop working for breaks etc. There is always a risk of human error within any task that is carried out. A good AI system will avoid mistakes and 'human error'.

Quality data analytics also benefits from the use of AI as it can be programmed to process vast amounts of big data which a business can use to make constructive and effective business decisions.

AR and VR

AR

Augmented reality is a view of the real world with computer-generated input (for example, graphics, audio visual or text) superimposed onto it. In essence, the AR content augments (enhances or adds to) the real-world scene that is viewed by the user. AR is an immersive visual experience and is commonly used to provide the user with information or guidance, allowing them to visualise things that they may not have access to or be able to imagine under normal circumstances.

IKEA, for example, provides an immersive AR experience through their app where a customer can select items such as furniture and 'place' them in their home. This allows the customer to visualise what the item will look like and where it can be placed, as well as how the colours and styles do or do not complement their current decor. This is a digital version of 'try before you buy'; a customer is able to see the visual impact of the item within their own home surroundings.

AR has also been used in oil and gas exploration to help engineers to navigate their way through complex equipment to carry out repairs, upgrades and/ or problem-solve faults. They can view the piece of equipment through an AR headset and are provided with instructions and images that can interact with and work through in order to identify a problem, dismantle the equipment to fit a new part or replace the item altogether.

Because the AR experience is immersive, the user feels part of the activity that is taking place. Firefighters are trained using AR for tackling incidents in dangerous situations, for example on oil refineries. They will train in empty oil tanks and around pipelines, wearing specialised AR glasses within their fire fighting helmets. As they navigate through the plant, they are presented with the types of potential hazards that they could come across while tackling the incident in real life. They are then required to interact with the AR app and make safety decisions. It is a safe method to train firefighters without the need for them to be in physical danger while training.

Within the medical profession, AR is used to help patients understand the prognosis of their ailments by overlaying augmented visualisation or annotation over things such as x-rays of the patient. They also use AR to tell the story of how a patient's operation will be performed.

VR

VR is where the user is totally immersed in the virtual world with no sight of the real world. This is facilitated using special VR headsets/glasses that block out any sighting of the real world. Unlike AR where the user interacts with the augmented real world, with VR the user can only interact with the virtual world.

VR is a popular training tool for the airline industry (training pilots to fly certain types of aircraft) and firefighters (to navigate their way through buildings full of smoke and fire without putting their lives in danger). The military also use VR for training as they can 'place' their military personnel in a wide variety of locations so that they can understand the environment in which they will be working and how to adapt to different situations.

VR is popular because it allows people to be trained on how to deal with what can be quite dangerous situations without having to be placed in actual danger. In addition, it can save time and money because people are not necessarily required to travel to a specific location. Imagine the cost of transporting an army battalion to the other side of the country or even to another country with all the equipment.

Research

VR and AR can have positive and negative impacts on society, individuals and organisations. Research the use of AR and VR within the tourism sector and create a presentation with speaker notes explaining the positive and negative impacts.

Test yourself

1 Identify the four pillars of IoT.
2 Explain the term 'deep learning'.
3 Describe how AI impacts society.
4 Explain the difference between AR and VR.
5 Identify one example of how VR is used by the military.

Skills practice

A school wants to implement the use of VR for lessons to motivate the students and to make the lessons more engaging. The lessons being considered initially are geography, history and the sciences.

The board of governors of the school has asked you to put together a business case to support the use of this innovative digital technology for them to consider. You have been asked to include:

▶ the moral and ethical issues that will be raised and how these challenges can be overcome
▶ the culture within the school that needs to be established, how it can be communicated in a positive way and how it can be sustained
▶ guidance for teachers on how they can use observational techniques to inform situational awareness of staff and students
▶ an explanation of what VR is and how it can be used for lessons.

Assessment practice

1 Discuss the positive and negative impacts of the IOT on society.
2 Explain the term 'autonomous operation'.
3 Identify three types of workers who are protected by law if they are whistleblowers.
4 Explain the term 'open-source'.
5 Discuss how organisations respond to moral and ethical issues when strategically planning for the design and development of digital systems.
6 Explain the difference between ML and deep learning.
7 Discuss the challenges and impacts of AI on society and individuals.
8 Describe two considerations when implementing change.
9 Identify three potential impacts that AI can have on businesses.
10 Explain the term 'inclusive workplace'.

Content area 4: Legislation and regulatory requirements

In this content area you will learn about some of the legislation that applies to the digital sector. The legislation covers a range of issues from the safe use of digital systems to how data and information should be processed. You will learn about the implications of international law and how this should be considered by those working in a range of job roles in the digital sector.

You will also learn how digital technologies can be used to monitor a workplace. The monitoring of a workplace can have both positive and negative effects on employees, and these will be considered.

In this content area you will learn about the codes of conduct that are produced by professional bodies for

their members. You will also learn about the codes of practice, the implications of their implementation and the differences between these and a code of conduct. Many employers and education providers will require an AUP to be signed. You will learn about the contents of these and the role of an AUP within a workplace or educational setting.

It is important that IT systems and digital devices are accessible to all users. You will learn about the different organisations that provide guidelines and standards to ensure that all users can access the data and information they require.

Learning outcomes

In this content area you will learn about:
4.1   Legislation
4.2   Guidelines and codes of conduct

4.1 Legislation

Industry tip

The details of each piece of legislation/Act were correct when this book was published. During your study for this course, you must make sure that you know about and understand the most up-to-date versions of each piece of legislation, including any changes or additional pieces of legislation that are relevant to the design, development and use of digital.

4.1.1 Health and safety when working with computers

The Health and Safety at Work Act provides guidance to employers and employees when working with computer systems. The Act also defines actions that an employer should take to protect employees who work with computers in their job.

Almost everyone, not just employees and employers, has a duty under the Health and Safety at Work Act to work and behave safely. The Act also makes it illegal to act recklessly, or intentionally act in such a way as to endanger yourself or others. Employees must take reasonable care for their own and others' safety and co-operate with their employers in doing so.

The Act applies to people using computers for their work but not necessarily to those using them at home, unless the employee works at home.

Display screen regulations

The main law covering the use of computer equipment is the Health and Safety (Display Screen Equipment) Regulations, section 4.1.1. The **DSE** regulations apply to any form of display screen. This means that employees working with PCs, laptops, tablets and smartphones are all covered by these regulations.

Key term

DSE: Display Screen Equipment

The DSE regulations cover employees who work with DSE for an hour, or more, during their working hours.

The regulations state that employers, for example a business, school or college, must complete four tasks to ensure the safety of their employees. These are:

1   Analyse workstations and assess and reduce risks

Employers need to check that the computer equipment and the area around it is safe. If any risks are found during the assessment of the workstation and surrounding area, then action needs to be taken to make it safe.

The assessment should also cover the job role being carried out and any special requirements of the staff member to carry out that job role.

Employers, and employees working from home, should complete a DSE workstation assessment. A DSE assessment should also be completed when:

▶ a new workstation is set up
▶ a new employee starts work

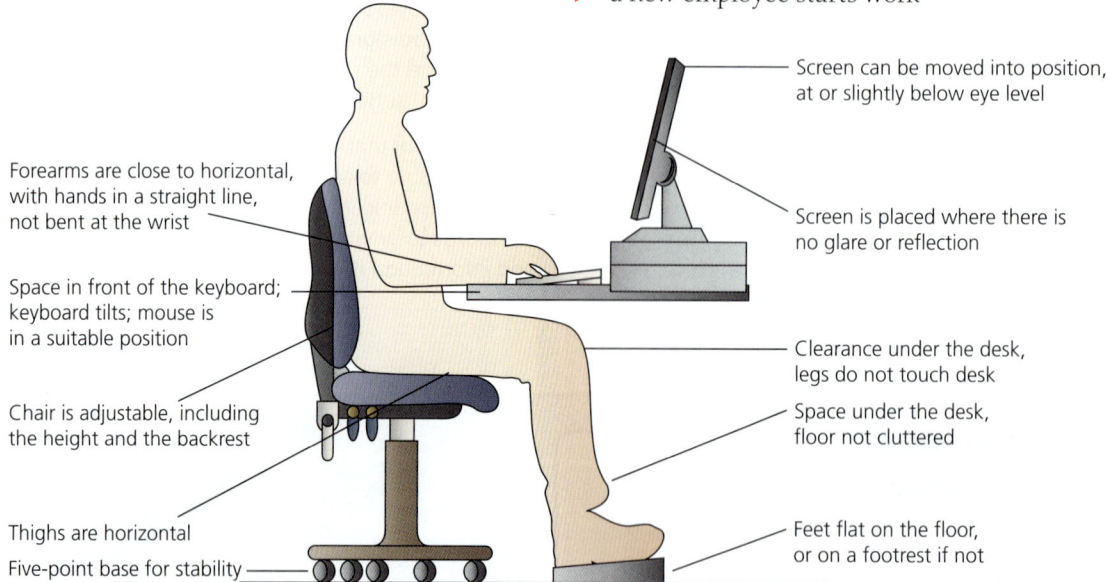

Forearms are close to horizontal, with hands in a straight line, not bent at the wrist

Space in front of the keyboard; keyboard tilts; mouse is in a suitable position

Chair is adjustable, including the height and the backrest

Thighs are horizontal

Five-point base for stability

Screen can be moved into position, at or slightly below eye level

Screen is placed where there is no glare or reflection

Clearance under the desk, legs do not touch desk

Space under the desk, floor not cluttered

Feet flat on the floor, or on a footrest if not

▲ Figure 4.1 The correct and safe arrangement of a workstation

- a change is made to an existing workstation or the way it is used
- employees complain of pain or discomfort.

Employers must ensure that workstations meet the minimum requirements including:

- providing adjustable chairs and suitable lighting
- providing tilt and swivel monitors
- ensuring that the workstation has sufficient space for the keyboard, monitor and any paperwork.

2 Plan work so that there are breaks or changes of activity

Employees should not be expected to work at DSE all day. Regular breaks should be provided or a change in the activity that the employees are carrying out. However, the regulations do not say how long or how frequent the breaks should be. In many workplaces a break from working at DSE may occur naturally, for example meetings in person. However, with the increase in employees working from home, this may not be possible as many meetings are held online.

In this case, employees should take responsibility for planning their own work day. The regulations do not define when, and for how long, breaks away from DSE should be taken, but it is better to have a 5–10 minute break every hour rather than the traditional tea break of 20 minutes and lunch break of 1 hour.

During a break, employees should be advised that they should move around, stretch and focus their eyes on anything except DSE.

3 Arrange and pay for eye tests and glasses (if special ones are needed)

Working with DSE does not cause permanent damage to eyes. But if DSE is used for a long time without a break it can lead to:

- tired eyes
- discomfort
- temporary short-sightedness
- headaches.

Employees of a business, who work with DSE, can ask for eye tests to be arranged and paid for by their employer. The eye tests can be repeated as advised by the optician – the employer will have to keep paying for these. The business will only have to pay for the glasses if the glasses are only needed for work.

4 Provide health and safety (Health and safety) training and information

Employers must provide training to make sure that their employees can use their computer equipment and workstations correctly. The training they provide could include how employees can use the equipment to minimise risks to their health. The employees should also provide information to their employees about Health and safety when using screen equipment and the steps that have been taken to minimise the risks.

Research

Investigate the Health and Safety Policy, which applies to DSE users, at your workplace or centre. Write down the main points of the policy and summarise how the policy affects the DSE users.

General working environment

The general working environment not only covers those employees working with DSE but all job roles within the business. The employer and employee both have responsibility for the working environment. The working environment should:

- have the appropriate workplace facilities including the right number of toilets and washbasins, drinking water, and somewhere to rest and eat meals
- be a healthy working environment
- be a safe workplace.

The Health and Safety Executive (HSE) states that a **healthy working environment** is one which has:

- good ventilation including a supply of fresh, clean air from outside or a well-maintained air conditioning system
- a reasonable working temperature so it is comfortable to work (usually at least 16°C, or 13°C for strenuous work, unless other laws require lower temperatures)
- lighting suitable for the work being carried out
- enough room space, suitable workstations and seating
- appropriate waste containers for recyclable and non-recyclable waste.

And a **safe workplace** is one which has:

- maintained buildings and work equipment
- floors and traffic routes kept free of obstructions
- windows that can be opened and cleaned safely
- any glass or transparent doors or walls protected or made of safety material.

Every workplace must have a Health and Safety Policy. This will cover areas such as (depending on the nature of the business):

▶ statement of intent – including the management of Health and safety

▶ responsibilities for Health and safety – the names of the responsible people and the specific area they have responsibility for

▶ arrangements for Health and safety – the practical arrangements including risk assessment, Health and safety training based on the employees' job roles and using safety equipment.

Activity

Create a Health and Safety Policy for your workplace or centre. Consider all the job roles that are carried out and the appropriate Health and safety procedures.

Where employees are using DSE, the **employer** has a responsibility to:

▶ provide adjustable chairs and suitable lighting

▶ provide tilt and swivel monitors

▶ ensure that a workstation has sufficient space for the keyboard, monitor and any paperwork.

When working with DSE it is the responsibility of the **employee** to:

▶ check their screen is well positioned and properly adjusted

▶ make sure lighting conditions are suitable

▶ take regular breaks from screen work.

Possible risks and prevention

By creating a Health and Safety Policy and carrying out risk assessment, many of the possible risks in a workplace will have been identified. However, due to the constant use of equipment, buildings and facilities, risks may occur. It is the responsibility of both employers and employees to keep their working environment safe.

This means that employees have a duty to report to the appropriate Health and safety responsible person any issues or accidents as soon as they have occurred. It is then the employer's responsibility to rectify the issue and/or report the accident to the Health and Safety Executive (HSE).

If an issue is reported then the employers may use signs to warn their employees of the hazard or fence off the area. It is the responsibility of the employees to take note of these signs and restricted area, and to conform with these restrictions.

Test yourself

1 Identify two tasks that an employer must complete to ensure the safety of their employees.

2 When should a DSE assessment be carried out?

3 What is meant by a healthy working environment?

4 What health issues could occur when working with DSE?

5 Identify two responsibilities of an employee when using DSE.

4.1.2 The Data Protection Act 2018/ General Data Protection Regulation

The Data Protection Act 2018 (DPA) attempts to control how personal data and information is used by organisations, businesses and the UK Government. The Act also seeks to empower individuals to take control of their **personal data** and to support organisations with their lawful processing of personal data.

The DPA is the UK's implementation of the General Data Protection Regulation (GDPR). The DPA does not bring the GDPR into UK law. The GDPR has had a direct impact on EU member states since 25 May 2018, which means the GDPR is already part of UK law. When the UK left the EU, the GDPR was amended and converted into UK law under the European Union (Withdrawal) Act 2018.

The Act provides an update to the DPA 1998 and became law on 25 May 2018. The Information Commissioner at the time, Elizabeth Denham, stated that:

'The previous Data Protection Act, passed a generation ago, failed to account for today's internet and digital technologies, social media and big data. The new Act updates data protection laws in the UK, provides tools and strengthens rights to allow people to take back control of their personal data.'

(ICO, 2019)

Key term

Personal data: any information relating to an identified or identifiable living individual.

The DPA:

> 'is a complete data protection system, so as well as governing general data covered by the GDPR, it covers all other general data, law enforcement data and national security data. The DPA includes a number of agreed changes to the GDPR to make it work for the benefit of the UK in areas such as academic research, financial services and child protection.'

(DCMS, 2018)

The principles of the act

The DPA defines principles. These can be broadly equated to the principles of the GDPR.

DPA 2018 Principles	GDPR Principles
Used fairly, lawfully and transparently	Lawfulness, fairness and transparency
Used for specified, explicit purposes	Purpose limitation
Used in a way that is adequate, relevant and limited to only what is necessary	Data minimisation
Accurate and, where necessary, kept up to date	Accuracy
Kept for no longer than is necessary	Storage limitation
Handled in a way that ensures appropriate security, including protection against unlawful or unauthorised processing, access, loss, destruction or damage	Integrity and confidentiality (security) / Accountability

▲ Table 4.1 Principles of the DPA 2018 and GDPR

The DPA also provides legal protection for sensitive information and data including:
- race
- ethnic background
- political opinions
- religious beliefs
- trade union membership
- genetics
- biometrics (where used for identification)
- health
- sex life or orientation.

> Some of the sensitive data and information is included in the Equality Act which is covered in section 4.1.4 of this content area.

Data subject rights

Under the DPA the **data subject** has rights. These include being able to find out what data is held or stored about them.

Other rights a data subject has under the DPA include the right to:
- be informed about how the data is being used
- access personal data
- have incorrect data updated
- have data erased
- stop or restrict the processing of the data
- data portability (allowing the data subject to get and reuse the data for different services)
- object to how the data is processed in certain circumstances.

A data subject also has rights when an organisation is using personal data for:
- automated decision-making processes (without human involvement)
- profiling, for example to predict behaviour or interests.

> **Key term**
>
> **Data subject:** the person the data is being held about.

One of the principles is that

> 'data and information is handled in a way that ensures appropriate security, including protection against unlawful or unauthorised processing, access, loss, destruction or damage'.

(DCMS, 2018)

This means that the data/information should be protected by some form of security. Effective data protection:

> 'relies on IT systems being protected from malicious intent. In implementing the GDPR standards, the Act requires organisations that handle personal data to evaluate the risks of processing such data and implement appropriate measures to mitigate those risks. These measures can include cyber-security controls.'

(DCMS, 2018)

> Unsubscribe | Forward to a friend
>
> If you'd like to speak to a product specialist, visit our website to find someone in your area.

▲ Figure 4.2 An unsubscribe link in a marketing email

> Security of digital systems, data and information is covered in Content area 8.

Research

Investigate each of the DPA principles. Create a digital communication providing an explanation of how a UK-based business can comply with each one.

Marketing consent

Under the DPA and GDPR consent must be given before, for example, a business, organisation or charity sends a marketing message. The consent must be knowingly and freely given and must be clear and specific. This is also covered under the current version of the Privacy and Electronic Communications Regulations (PECR) summarised below.

The consent needs to be given for the organisation and the type of marketing communication to be used such as a phone call, automated call, email or text. The consent should be a positive action, for example ticking a box, clicking an icon or sending an email.

The person must fully understand that they are providing consent. It is not acceptable to only provide information about marketing as part of terms and conditions or a privacy policy as these can be hard to find, difficult to understand or rarely read.

The clearest way to obtain consent is to ask the person to tick a box confirming they are happy to receive marketing calls, texts or emails. This is called **opt in**.

It is important that clear records of what has been consented to are kept, including when and how the consent was given. This means that compliance can be proved in case of a complaint.

Consent can be withdrawn at any time, so it must be easy for someone to do this. This is called **opt out**. For example, a link at the bottom of a marketing email that provides a link to unsubscribe.

Some organisations provide opt-in boxes that are automatically pre-ticked. However, the GDPR is clear that pre-ticked boxes do not give valid consent.

Key terms

Opt in: means a person has to take a specific positive step, for example tick a box, send an email or click a button or icon, to say they consent to receiving marketing.

Opt out: means a person must take a positive step to refuse or unsubscribe from marketing.

ICO: Information Commissioner's Office.

Enforcement

Those who do not meet the requirements will face penalties and other restrictions. There are two tiers of penalty that can be issued by the **ICO** – the higher maximum and the standard maximum.

The higher maximum penalty applies to any failure to comply with the core data protection principles or an individual's rights as stated by the DPA. It also applies in the instance of data transfers to third counties. The penalty is £17.5 million or 4% of the total annual worldwide turnover in the preceding financial year, whichever is higher.

For all other infringements (for example of the DPA's administrative requirements) the standard maximum penalty applies. This is a penalty of £8.7 million or 2% of the total annual worldwide turnover in the preceding financial year, whichever is higher.

4.1.3 Computer Misuse Act

The Computer Misuse Act (CMA) was brought into UK legislation in 1990 and has seen many amendments over the years.

The original focus of the CMA was to criminalise the act of accessing or modifying data stored on a computer system without appropriate consent or permission. In 1990 the use of computer devices and networks was limited but was increasing. This increased use led to more data being stored on computer devices and networks.

The Act was developed following a high-profile case in 1987, Regina v Gold and Schifreen. Gold and Schifreen were hackers who managed to gain remote access to the British Telecom (BT) Prestel Service at a trade show. They used the authentic access details of a BT engineer to carry out the remote access attack. They acquired the access details by shoulder surfing.

The definition of access or modification was initially limited as the number of methods by which the Act could be broken was very limited. However, with the increased use of computer devices, networks and the internet, there have been several amendments to the CMA. The increased use and storage of data has led to an ever-increasing range of threats and possible areas for harm. The act of preparation for a cyber attack and a range of methods of attack have now been included.

It is interesting that the CMA does not provide a definition of a computer. This is justified because if definitions were included they could very quickly become out of date due to the ever-changing innovations in technology.

The CMA enables the justice system to determine the definition based on the case that is in front of them. However, Lord Hoffman defined a computer as:

'a device for storing, processing and retrieving information.'

(DPP v McKeown, DPP v Jones, 1997)

Key term

DPP: Director of Public Prosecutions.

The three original sections of the CMA covered:

Section	Offence	Penalty
1	Unauthorised access to computer material	A fine up to a maximum of £5,000 and/or a maximum six months in prison
2	Unauthorised access to computer materials with intent to commit a further crime	An unlimited fine and/or maximum five years in prison
3	Unauthorised modification of data	An unlimited fine and/or maximum five years in prison

▲ Table 4.2 The original sections of the CMA

A further two sections were included in the CMA in the 2000s. The Police and Justice Act of 2006 led to section 3A being included in the CMA. The introduction of the Serious Crime Act of 2015 led to section 3ZA being added to the CMA.

Section	Offence	Penalty
3A	Making, supplying or obtaining any articles for use in a malicious act using a computer	An unlimited fine and up to five years in prison
3ZA	Unauthorised acts causing, or creating risk of, serious damage	An unlimited fine and/or up to 14 years in prison, unless the offence caused or created significant risk to human welfare or national security – in which case there is a maximum life sentence in prison

▲ Table 4.3 Two further sections added to the CMA

Section	Examples of offences
1	Hacking Without them knowing, you watched your friend put their password into their phone. You then used it to gain access to their phone and download their photos.
2	Obtaining the unauthorised access with the intention of committing theft, such as by diverting funds, which are in the course of an electronic funds transfer, to the defendant's own bank account, or to the bank account of an accomplice; or where the defendant gained unauthorised access to sensitive information held on computer with a view to blackmailing the person to whom that information related. Without their permission, you accessed your friend's smartphone, obtaining their bank login details, so you could transfer money from their account.
3	You learned from a YouTube video how to use a webstresser or booter tool to perform a denial of service (DoS) attack against a friend, knocking them off an online game. You did this simply intending to win a game.
3A	You downloaded a program which was able to take remote control of a friend's computer without their knowledge. You didn't get a chance to use it before you were caught. This offence covers the possession of 'malware' but also legitimate software for which you had the intent of using it to commit an offence.
3ZA	The most serious cyber attacks, for example those on essential systems controlling power supply, communications, food or fuel distribution. These could have a significant impact, resulting in loss of life, serious illness or injury, severe social disruption or serious damage to the economy, the environment or national security.

▲ Table 4.4 Example offences for the CMA sections

Examples of offences

Table 4.4 shows examples of offences for each of the sections of the CMA.

Hacking

It can be seen that section 1 of the CMA refers to hacking.

Hacking means finding out weaknesses in an established system and exploiting them. A computer hacker is a person who finds out weaknesses in a computer system to gain unauthorised access. A hacker may be motivated by a multitude of reasons, such as profit, protest or challenge.

There are three main types of hacking that can take place. These are white hat, grey hat and black hat.

▶ **White hat** hacking is where the hacker is given permission to hack into systems to identify any loopholes or vulnerabilities. As this type of hacking is done with the permission of the computer system owner, it does not break any of the legislation that relates to hacking.

▶ **Grey hat** hacking is where the hacker hacks into computer systems for fun or to troll but does not have malicious intent towards the computer system. If a grey hat hacker finds a weakness then they may offer to fix the vulnerability – but for a fee. Grey hat hackers can also manipulate the rankings of websites when a search is done on a search engine.

▶ **Black hat** hacking is where the hacker hacks into a computer system with malicious intent. This intent can include stealing, exploiting the data stolen or seen, and selling on the data. Black hat hackers carry out illegal hacking activities and can be prosecuted under UK IT legislation.

Examples of hacking attacks

There have been many high-profile examples of hacking. Most of these have included some form of data breach. Some examples are:

Adobe

In October 2013 Adobe reported that their systems had been hacked. It was reported that 38 million usernames and encrypted passwords had been stolen by the hackers. In addition to this data breach, Adobe reported that the hack had exposed an undisclosed number of customer names, IDs, passwords and debit and credit card details.

eBay

In May 2014 eBay reported that a hacking attack had exposed account details for all its users, reported to be 145 million people. The account details included names, addresses, dates of birth and encrypted passwords. eBay confirmed that hackers had used the credentials of three corporate employees to access its network and had complete access for 229 days.

British Airways

Between 22:58 on 21 August 2018 and 21:45 on 5 September 2018 British Airways (BA) suffered a hacking attack which led to a data breach of BA customers' data. The data breach was limited to those customers who had booked flights and holidays during that time through the BA website or mobile app. About 380,000 customers were found to have been affected. The data that was stolen related to payment details. BA reported that no travel or passport data was stolen during the attack.

> **Research**
>
> Research high-profile hacking attacks on businesses. Develop a presentation showing the results of your findings. Present your findings to your group.

Threats

There are many different threats to data and information. Some threats can also be targeted at physical computer equipment.

Cyber-security attackers, including hackers, can hijack a computer system which could lead to data and information being stolen. This can have huge impacts on a business or an individual. As more businesses and people use the internet for, for example financial transactions, security of this data has to be increased to ensure that information and data does not fall into 'the wrong hands'.

Hackers use a range of methods and threats to gain information and data.

These methods and threats can be split into three broad categories:

▶ DDoS
▶ **malware** including viruses
▶ social engineering.

> **Key term**
>
> **Malware:** malicious software.

DDoS

This is an attempt to make a computer or network system unavailable to its users by flooding it with network traffic. A DDoS is usually focused on preventing an internet website or service from either functioning efficiently, or at all, either temporarily or indefinitely. DDoS attacks usually target websites or services hosted on high-profile web servers, such as banks, payment websites, for example Google Pay or PayPal, and mobile phone companies.

Malware

This is installed on a computer system. Some malware collect information about users without their knowledge whilst other will disrupt the digital system it has infected. There are many types of malware, but the main types are adware, bot, bug, ransomware, rootkit, spyware, Trojan horse, virus and worm.

It is simple to protect against many forms of malware by installing, running and keeping up-to-date security software. It is also not wise to open any suspicious files or click on any links in emails.

Table 4.5 shows different types of malware, why they are used, how they work and how to mitigate against them.

Type of malware	Why they are used	How they work	How to mitigate
Adware	Adware generates revenue for its author.	Adware is also known as advertising-supported software. This is any software package which automatically shows adverts, such as a pop-up. It may also be in the UI of a software package or on an installation screen. Adware, by itself, is harmless; however, some adware may include spyware such as key loggers.	Install, run and keep updated a security software package. Do not open any files from an unknown source. Do not click any links in an email.
Bot/botnet	Bots take control of a computer system.	A bot is a type of malware that allows a cyber-security attacker to take control of a computer system that has been infected without the user's knowledge. It can result in a botnet which is an interconnected network of infected computer systems.	Install run and keep updated a security software package. Do not open any files from an unknown source. Do not click any links in an email.

Type of malware	Why they are used	How they work	How to mitigate
Bug	Bugs are connected to software and are flaws that produce an unwanted outcome.	Bugs are usually the result of human error during the coding of the software. Most bugs can be fixed by the software creator issuing a fix or patch. Security bugs are the most severe type and can allow cyber attackers to bypass user authentication, override access privileges or steal data.	Check for and install any patches that are released from software vendors.
Ransomware	Ransomware holds a computer system captive and demands a ransom, usually money, to release it.	Ransomware can restrict user access to the computer system by encrypting files or locking down the computer system. A message is usually displayed to force the user to pay so that the restrictions can be lifted and the user has access to the data/computer system. It is spread like a worm and can be started by downloading an infected file or by a vulnerability in the computer system.	Do not open any files from an unknown source. Do not click any links in an email. Install, run and update security software.
Rootkit	A rootkit is designed to remotely access or control a computer system without being detected by the security software or the users.	When a rootkit has been installed it can enable a cyber attacker to remotely access files, access/steal data and information, modify software configurations or control the computer system as part of a botnet.	Rootkits are difficult to detect as they are not usually detected by security software. Software updates, keeping security software up to date and not downloading suspicious files are the only ways of trying to avoid a rootkit from being installed.
Spyware	Spyware can collect data from an infected computer, including personal information like websites visited, user logins and financial information.	Spyware is usually hidden from a user and can be difficult to detect. Spyware is often secretly installed on a user's personal computer without their knowledge. However, some spyware such as key loggers may be installed to intentionally monitor users. Spyware can also install additional software or redirect web browsers to different websites. Some spyware can change computer settings which could lead to slow internet connection speeds or changes in web browser settings.	Do not open any files from an unknown source. Do not click any links in an email. Install, run and update security software.
Trojan horse	A Trojan horse is a standalone malicious program designed to give full control of a PC infected with a Trojan to another PC.	Trojans often appear to be something which is wanted or needed by the user of a PC. They can be hidden in valid programs and software. Trojan horses can make copies of themselves, steal information or harm their host computer systems.	Do not open any files from an unknown source. Do not click any links in an email. Install, run and update security software.
Virus	A virus attempts to make a computer system unreliable.	A virus is a computer program that replicates itself and spreads from computer to computer. Viruses can increase their chances of spreading to other computers by infecting files on a network file system or a file system that is accessed by other computers.	Do not open any files from an unknown source. Do not click any links in an email. Install, run and update security software.
Worm	A standalone computer program that replicates itself so it can spread to other computers.	A worm can use a computer network to spread. Unlike a computer virus, it does not need to attach itself to an existing program. Worms almost always cause some harm to a network, even if only by consuming bandwidth.	Do not open any files from an unknown source. Do not click any links in an email. Install, run and update security software.

▲ Table 4.5 The different types of malware, why they are used, how they work and how to mitigate them

Social engineering

Social engineering can also take many forms. It is the art of manipulating people so that confidential information can be found out. The most common ones are phishing, pretexting, baiting, quid pro quo, tailgating/piggybacking and shoulder surfing.

Table 4.6 shows the different types of social engineering, why they are used and how they work. Many victims of social engineering are people who are not tech-savvy.

Type of social engineering	Why they are used	How they work
Baiting	Baiting tries to trick the victims to give the cyber criminals the information they need.	Baiting is very similar to phishing. The cyber criminals will make a promise of an item or goods to get the information they need. An example would be to promise free downloads of films or music in return for log-in details.
Phishing	Phishing tries to get users to input, for example, their credit or debit card numbers and security details, or log-in details into a fake website.	Phishing uses a fake website which looks identical to the real one. The most common targets for phishing are bank, building society and insurance websites. The attackers send out emails or text messages which pretend to be from, for example, your bank. A link is contained in the email which you are asked to click on. This link takes the user to a fake website.
Pharming	Pharming is a cyber-security attack that tries to redirect users from a genuine website to a fake one. This is done without the knowledge of the user.	Pharming is very similar to phishing in that both use fraudulent websites. The main difference is that a phishing attack will use fake or hoax emails while pharming attacks very rarely use this type of tactic.
Pretexting	Pretexting is when a cyber criminal lies to get data or information.	Pretexting usually involves a scam where the criminal pretends to need the information to confirm the identity of the person they are talking to.
Quid pro quo	Quid pro quo tries to disable the anti-virus software so that software updates, usually malware, can be installed to gain access to a computer system.	Quid pro quo is very similar to baiting; however, the promise is that of a service rather than goods. A common method of quid pro quo is a telephone call from a fake IT service provider. These people offer IT assistance in fixing problems which usually don't exist.
Scareware	Scareware is a malicious computer program.	Scareware is designed to trick a user into buying and downloading unnecessary and potentially dangerous software, such as fake anti-virus protection.
Shoulder surfing	Shoulder surfing aims to steal data and information.	Shoulder surfing is when a person's private and confidential information is seen. For example, when using this method the attacker may stand very close to someone using a cash machine to see their PIN. This method is very effective in crowded places when a person uses a smartphone or mobile device and log-in details can be seen.
Smishing	Smishing is a form of phishing and is the fraudulent practice of sending text messages.	Smishing is when someone tries to trick you into giving them your private information, such as passwords or credit card numbers, via a text or SMS message which pretends to be from a reputable company.
Tailgating/ piggybacking	Tailgating/piggybacking is used to try to gain access to a secure building or room.	This type of attacker takes the form of someone who does not have authority to enter a building or room, following someone who does through the doors. The most common type is that the attacker pretends to be a delivery driver and asks an authorised person to hold the door.
Vishing	Vishing is making phone calls or leaving voice messages to try and trick the recipient.	The calls and messages pretend to be from reputable companies to try and trick people into revealing personal information, such as bank details and credit card numbers.

▲ Table 4.6 The different types of social engineering, why they are used and how they work

Further details about threats to digital systems can be found in Content area 8.

Research

Investigate the Glossary of Terms on the Equality and Human Rights Commission website. Can you think of any terms that are not included?

Test yourself

1 Identify two offences defined under the original CMA.
2 What is the penalty for a breach of section 3ZA?
3 What is grey hat hacking?
4 How does ransomware work?
5 Why is phishing used?

4.1.4 Equality Act

The Equality Act 2010 became law in October 2010. The aim of the Act was to provide equality of opportunity for all people. The Act protects people from any form of discrimination in society and the workplace.

The Act replaced and combined over 116 different pieces of legislation related to discrimination, harassment and victimisation. This made the legislation easier to understand and apply. The Equality Act defines the different ways in which it is unlawful to treat people.

The Act provides a legal framework to protect the rights of individuals and advance equality of opportunity for all. It provides the UK with an anti-discrimination law which protects individuals from unfair treatment and promotes a fair and more equal society.

The Equality Act protects everyone in all situations including in the workplace. This is because the Act protects people against discrimination, harassment and victimisation because of the protected characteristics that we all have. It is those protected characteristics that makes us all individual and unique.

As with all legislation, the Equality Act contains specific terms. The Equality and Human Rights Commission has a Glossary of Terms on their website.

Types of discrimination (protected characteristics)

The Equality Act defines nine protected characteristics. These are:

▶ age
▶ disability
▶ gender reassignment
▶ marriage and civil partnership
▶ pregnancy and maternity
▶ race
▶ religion or belief
▶ sex
▶ sexual orientation.

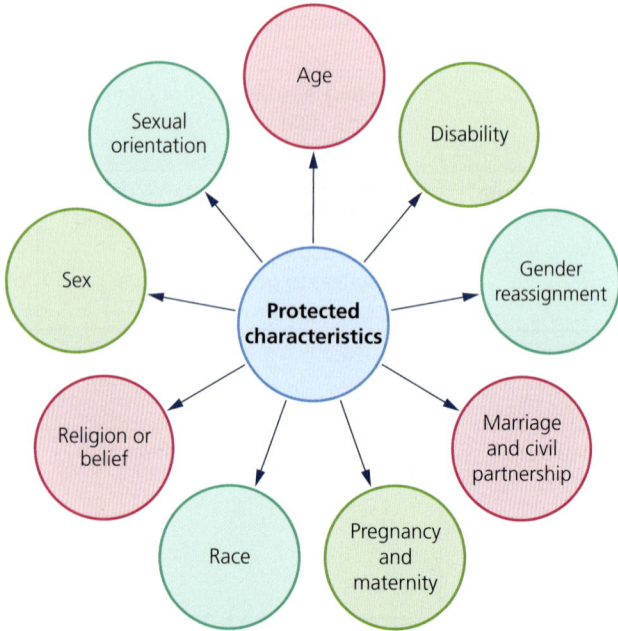

▲ Figure 4.3 The nine protected characteristics defined in the Equality Act

How individuals can be discriminated against

There are different types of discrimination that can be applied to each of the characteristics. The four main types are shown in Table 4.7.

Type	Meaning	Example
Direct discrimination	Treating one person worse than another person because of a protected characteristic.	An older employee is not trained on a new software system as the employer feels that they will be retiring soon so to train the employee would be a waste of time and money.
Indirect discrimination	This can happen when an organisation puts a rule or a policy or a way of doing things in place which has a worse impact on someone with a protected characteristic than someone without one.	A new digital system is introduced which uses a touch screen. The system does not support assistive technology so employees who need assistive technology are moved to a lower paid job role.
Harassment	This means people cannot treat you in a way that violates your dignity, or creates a hostile, degrading, humiliating or offensive environment.	Employees who are unable to complete their job role due to a new digital system are the victims of derogatory name calling in the workplace.
Victimisation	This means people cannot treat you unfairly if you are making a complaint of discrimination under the Equality Act or if you are supporting someone else who is doing so.	Employees who cannot use a new digital system which uses a touch screen make a complaint to their employer. These employees are the first group to be made redundant when a restructuring takes place.

▲ Table 4.7 The four main types of discrimination

Characteristic	Types of discrimination
Age	Direct discrimination
	Harassment
	Indirect discrimination
	Victimisation
Disability	Direct discrimination
	Discrimination arising from disability
	Failure to make reasonable adjustments
	Harassment
	Indirect discrimination
	Victimisation
Gender reassignment	Direct discrimination
	Harassment
	Indirect discrimination
	Victimisation
Marriage and civil Partnership	Direct discrimination
	Indirect discrimination
	Victimisation
Pregnancy and maternity	Unfavourable treatment
	Victimisation
Race	Harassment
	Victimisation
Religion or belief	Direct discrimination
	Indirect discrimination
Sex	Direct discrimination
	Harassment
	Indirect discrimination
	Victimisation
Sexual orientation	Direct discrimination
	Harassment
	Indirect discrimination
	Victimisation

▲ Table 4.8 The types of discrimination that can be applied to each protected characteristic

Each characteristic is covered by different types of discrimination. Most of the characteristics are covered by the four main types, as shown in Table 4.7. However, you will notice that some characteristics are covered by other types of discrimination.

Table 4.8 shows the types of discrimination that can be applied to each characteristic.

Research

Select one of the protected characteristics. Investigate examples of discrimination for your protected characteristic. Create a digital communication detailing what is meant by the characteristic and showing examples of discrimination. Discuss your findings with the rest of your group.

Where individuals are protected

You have already learned that people are protected in the workplace and in wider society. There is a defined range of situations where a person is protected from discrimination. These are:

'when you are in the workplace

when you use public services like healthcare (for example, visiting your doctor or local hospital) or education (for example, at your school or college)

when you use businesses and other organisations that provide services and goods (like shops, restaurants and cinemas)

when you use transport

when you join a club or association (for example, your local tennis club)

when you have contact with public bodies like your local council or government departments.'

(EHRC, 2020)

The aim of the Equality Act is to provide a framework for inclusivity in the workplace and society. However, there are some circumstances where being treated differently, in terms of the protected characteristics, is acceptable.

For example, an organisation providing help and advice for immigrants from India may specify that all their employees must be of Indian descent and speak one of the native Indian languages.

> ### Research
>
> For each protected characteristic investigate the circumstances where being treated differently to the Equality Act is acceptable. Make notes about your findings.

When to take action against discrimination

An individual could take action if they feel they have been discriminated against. The action could be to:
- complain to your employer either informally or using a formal grievance process
- ask for help and support from, for example, a trade union, equality organisation or the Citizens Advice. Someone from these organisations can be asked to act on your behalf. This is sometimes called a mediation process.
- begin Employment Tribunal proceedings.

There are organisations who can help if an action against discrimination is being considered. These organisations include professional bodies, trade unions, citizens advice and the dedicated equality advisory support service (EASS).

> ### Test yourself
>
> 1 How many pieces of legislation did the Equality Act combine and replace?
> 2 Identify three protected characteristics under the Equality Act.
> 3 What is meant by the term 'direct discrimination'?
> 4 What types of discrimination can be applied to the protected characteristic of gender reassignment?
> 5 Identify two situations where a person is protected from discrimination.

4.1.5 Intellectual Property Act 2014

The Intellectual Property Act (IPA) came into force on 1 October 2014. The aim of the Act was to streamline, simplify and strengthen design protection and patents.

Unregistered and registered designs

There are two types of design protection included in the IPA:
- Registered Design Right Protection
- Unregistered Design Right Protection.

A registered design right gives complete control over the design, while an unregistered design right only gives the ability to prevent others from copying the design. This means that the level of protection of an unregistered design right is lower than that of a registered design.

The IPA defines that the designer of a commissioned design will now own the design and not, as prior to this Act, the person commissioning the design. An exception to this is if the contract clearly states that the commissioner owns the designs. This change is specific to commissioned designs created by a self-employed or independent designer. If an employee creates a design as part of their job role then the employer owns the design.

This, however, only applies to those designs created on or after 1 October 2014 and applies to both UK registered and unregistered design rights. The Act cannot be retrospectively applied to designs created before this date.

The IPA includes a prior use clause. This means that if a design is being used prior to it being registered by someone else, they will be allowed to continue to use the design without any penalty. But if the design is changed then the IPA may have been broken.

It is now a criminal offence to intentionally copy a UK or EU registered design. To be illegal the copying must be without the agreement of the design owner and the offender must know that the design is registered.

Activity

Complete the table to show the differences between a registered and unregistered design.

	Registered design	Unregistered design
How a design is protected		
Type of protection		
How long the protection lasts		
Originality		
Protection rights		
Protection in other countries		

Patents

A patent can be granted to give the owner of a design a legal right to stop others from making, using or selling it for a specified number of years. Patents can be applied to all areas of technology, as long as they are new, involve an inventive step and can be applied to industry.

It used to be that a patent number had to be shown to define that the design had been patented. The IPA supplies details that allows the patent holder to provide an internet link to a website which gives details of the patent number. As with other details this only applies to designs patented after 1 October 2014.

Patents can be applied to most designs, including software processes and hardware.

Research

Investigate hardware and software patents. Make notes about your findings.

Test yourself

1 What are two types of design protection included in the IPA?
2 What is a prior use clause?
3 What is a patent?
4 How can a patent be shown after 1 October 2014?

4.1.6 The use of digital technologies for monitoring the workplace

Workplace monitoring enables an employer to track employee activities and then monitor employee engagement with work-related tasks. A business using employee monitoring on a digital device can measure productivity, track attendance, ensure security and collect proof of hours worked.

There are several different ways in which employees can be monitored. The most common use of monitoring is in relation to electronic communication.

Monitoring electronic communications

This includes the monitoring of:
▶ computer screens
▶ email
▶ internet and app use
▶ phone use.

The Telecommunications Regulations 2000 allow employers to monitor employees without the employee having given their consent first. Employers must clearly explain the amount of monitoring in the staff handbook or contract. These details may be also included in the AUP:
▶ if/how they are being monitored
▶ if personal emails and calls are not allowed
▶ the acceptable number of personal emails and phone calls.

Research

Research the acceptable purposes for monitoring employees as detailed in the Telecommunications Regulations 2000. Make notes about your findings.

Further details about an AUP can be found in section 4.2.5 of this content area.

There are several reasons an employer may want to monitor electronic communications. These include to:

▶ identify criminal activity

▶ check that employees are working to necessary standards

▶ check that employees are following the correct procedures

▶ investigate allegations of misconduct (for example if an other employee says that they are being bullied or harassed)

▶ see if there has been any mishandling of confidential information

▶ see that employees are not abusing work systems.

Most employers have internet access which employees use as part of their job role. It can be helpful for an employer to monitor the websites accessed during working hours. Internet monitors can be installed to detail which websites are being accessed. For example, if the internet monitor audit log shows details of shopping websites being accessed it can be assumed that the employees are not focusing on their job roles.

However, there are some job roles where access to websites such as social media are required. A business may have a social media manager who, as part of their job role, uses social media such as Facebook, Instagram and Twitter. Accessing and posting to these websites are the main part of the job role. So, the monitoring audit logs would show these websites being accessed, but are these being accessed as part of their job role or being accessed for personal use?

Websites can also be blocked. Many employers will restrict their employees from visiting websites with inappropriate content, while others may allow websites, like Twitter or YouTube, but only for a limited period of time.

Phone calls can be monitored. When a contact number for a business is dialled, it is quite common to hear an automated message stating that the call may be recorded for training or monitoring purposes.

This message means that the employer is recording all phone calls. This can cause a moral dilemma. If employees are allowed to make and receive personal calls then these too will be recorded.

Recording calls can help with training. If an employee gets constant positive feedback from customers then their calls can be listened to. By doing this, the calls can be used to see why customers provide positive feedback. The calls can also be used to train those employees who are either new or who do not receive as much positive feedback.

Those businesses that are only concerned about the misuse of phones can just record the numbers dialled and how much time is spent on the calls.

There are advantages and disadvantages to using monitoring and monitoring software in a workplace.

The **advantages** include:

▶ Employees can work flexible hours as monitoring can ensure all employees complete their required tasks.

▶ The most productive employees can be identified and rewarded by, for example, a promotion or a pay rise.

▶ Delivery drivers can be tracked to ensure their safety and that of delivery vehicles and contents.

The **disadvantages** include:

▶ Employees can feel that they are not trusted and that the monitoring is an invasion of privacy.

▶ Employee morale may reduce due to continual monitoring and lack of trust.

▶ Employee stress levels may increase which could lead to an increase in number of employees off sick.

Use of secret monitoring

Employers monitoring workplace and/or employees without telling them this is happening. Methods of secret monitoring can include hidden cameras or audio devices.

The use of secret monitoring of a workplace and/or employees is very rarely legal in the UK. Data protection legislation makes it illegal to secretly monitor certain areas within a workplace, for example staff bathrooms, unless the employer has strong suspicions that serious crime, for example drug dealing, is taking place in those areas.

Employers' monitoring policies

Many employers will include details about the workplace monitoring that is carried out in the contract of employment or the AUP. By including the monitoring details in either, or both, of these documents then it can be guaranteed that employees will have a full and working knowledge of the policy. So, if an employee is breaking the terms of the monitoring policy, then disciplinary processes can be implemented.

> **Further details about an AUP can be found in section 4.2.5 of this content area.**

Monitoring systems

There are many different systems that can be used to monitor a workplace and employees. These include:

▶ key logging
▶ video/audio surveillance
▶ Global Positioning System (GPS) vehicle tracking
▶ location tracking by access badge.

Key logging

A keylogger is a type of software that runs in the background of a digital system. The keylogger records every key press and may also record every mouse click or touchscreen user interaction.

A keylogger can record most actions a digital system user takes. This includes username and password input, emails written and any websites visited. It may also be that the keylogger can enable screenshots of an employee's screen to be recorded.

A keylogger can also record what programs or apps an employee has open at any one time. It is also possible to record and track any conversations including private chats, instant messaging or work-based chat facilities.

Video and audio surveillance

Video surveillance is usually carried out through the use of closed-circuit television (CCTV) cameras. When CCTV is operating the employer should, as with all other monitoring systems, ensure their employees are told about it and the reason for the use of the CCTV cameras. However, cameras should not be installed in areas of a workplace where people expect complete privacy, for example toilets and changing rooms.

Most workplaces covered by CCTV cameras have signs providing information about the location of the cameras. It is important that the signs are:

▶ clear, visible and readable
▶ show details of the purpose of the CCTV surveillance and who to contact
▶ include contact details, for example website address, telephone number or email address.

What must be considered by employers is that under the Data Protection Act, the CCTV cameras must only be used for the purpose provided and cannot be used for anything else. For example, if the reason given is to prevent theft, then the cameras must not be used to record employees and visitors coming into and going out of the workplace. The employer must also inform the ICO that they are using CCTV in their workplace.

The GDPR clarifies that CCTV footage is personal information. The GDPR includes a number of specific requirements about how personal information is stored and processed. Employers must also provide any CCTV footage if a subject access request (SAR) is made.

The DPA also details the status of CCTV footage and outlines requirements for the collection, processing and disclosure of CCTV data.

Many CCTV cameras can also record audio. But CCTV audio must not be recorded as this is not allowed other than under exceptional circumstances. The only exceptions to this rule include panic buttons in a taxi or monitoring carried out in a private area of a police custody room. However, if it is deemed necessary and justifiable then, as with other monitoring systems, all employees must be told that both video and sound are being captured by cameras.

> **Further details about the DPA and GDPR** can be found in section 4.1.2 of this content area.

GPS vehicle tracking

GPS is a method of working out exactly where something is. A GPS tracking system can be installed in vehicles to allow employers to see the exact location of any of their vehicles and the employees who are driving them.

GPS uses a network of satellites that continually transmit time and location messages. A GPS receiver, such as a smartphone, will use the signals from four different satellites to calculate its location.

When relying solely on satellite data the accuracy of GPS is within five metres. However, when extra signal sources are used such as using data from mobile phone towers the accuracy can be to one metre.

As with all monitoring systems transparency between employer and employees is important. So, the employer should tell those driving the vehicles that GPS tracking is installed.

There are advantages and disadvantages of using GPS tracking systems. The most important advantage is that of safety. GPS tracking allows employers to track where a vehicle is at any given time, so, if a vehicle is not moving, the driver can be contacted to check on their safety. If the vehicle has had a breakdown or an accident, the exact location can be used to send help and assistance.

One disadvantage of GPS tracking is that of a lack of a clear line of sight. Anything that blocks the signal can cause problems, with erratic data and inconsistent service. This can be an issue if the vehicle is in a remote location. Things that can block a line of sight include buildings, trees, tunnels and hills. However, if the vehicle is moving these problems tend to be temporary.

> **Research**
>
> Investigate the advantages and disadvantages of GPS vehicle tracking. Make notes about your findings.

Location tracking by access badge

It is very common for employees to use a security badge in the workplace. The badges can show details of the employee, often including a photograph, and are used to get access to a workplace or restricted access to different areas within the workplace.

Each badge contains a unique key which is allocated to a specific employee. It is this key that provides or declines access to offices or restricted areas. In the case of an emergency, it is easy to determine which employees are in the workplace and who has already left.

However, badges can be left at home, broken or lost. This leads to a cost to the employer who has to replace the badge. This could also lead to a security breach if the badge is found and used to carry out criminal activity.

Employees can be tracked wherever they are in the workplace. This can lead to a feeling that 'Big Brother' is watching. However, this enables the exact location of an employee to be shown which could save a life if, for example, a fire broke out.

Badges do come at a cost. They can be lost, or former employees can fail to return them before leaving the company. They can be cracked or left at home, in a car or in a jacket pocket. Many companies are replacing separate security badges with smartphone-based access apps. These apps provide the same access as a badge but are more robust, especially when it comes to location tracking services.

> **Test yourself**
>
> 1 Identify two electronic communication types that could be monitored.
> 2 Describe one advantage and one disadvantage of using monitoring in a workplace.
> 3 Identify two different methods of monitoring systems.
> 4 Identify one Act that relates to the use of CCTV cameras in the workplace.
> 5 Describe one disadvantage of using GPS vehicle tracking.

4.1.7 The role of legislation relating to international law and its importance when designing, developing and using digital systems

As has been discussed in this content area, co-operation between digital professionals including those who use digital systems is very important. This means that co-operation across country boundaries must be maintained. There are many pieces of legislation that cross country borders. As the transition period for the UK to leave the EU ended on 1 January 2021, amendments have been made to some of the EU legislation to bring these into UK legislation.

The design and development of digital systems is, in the main, covered by the guidelines provided by organisations such as the ISO, W3C, WCAG and IETF. These need to be considered at the design stage, even when they relate specifically to the use of digital systems. When designing digital systems, it is also very important to consider the specific local legislation of the country or area in which a product will be used, as these tend to differ.

> Further details about the International Organization for Standardization (ISO), World Wide Web Consortium (W3C), Web Content Accessibility Guidelines (WCAG) and the Internet Engineering Task Force (IETF) can be found in section 4.2.4 of this content area.

The international legislation relating to the use of digital systems include:
- European Convention on Human Rights (ECHR) – Article 8
- Electronic Communications Privacy Act (ECPA)
- Controlling the Assault of Non-Solicited Pornography and Marketing (CAN-SPAM).

ECHR – Article 8

Article 8 protects a person's right to respect for their correspondence, family life, home and private life. Correspondence includes emails, letters and phone calls. This basically means that a digital system cannot be used to 'snoop' on anyone.

There are, however, exceptions to this. For example, it is possible for covert surveillance to be carried out if there is a perceived threat to national security, for example a terrorist attack. It is also possible for surveillance to be carried out in the case of the most serious criminal activity.

The increased use of the internet has enabled new types of political activism, cultural exchanges and the exercise of human rights. However, some may argue that restrictions related to access to the internet and digital media, and attempts by governments to monitor online activities or e-communications, for example email, messaging and social media, interfere with the fundamental rights defined in the ECHR and possibly freedom of religion and belief, or the right to a fair trial.

The internet is largely unpoliced: this was a founding principle for Tim Berners-Lee. However, the increased use of the internet in everyday life has also led to a darker side of human nature. The internet has enabled the increased space for criminal behaviour and activity which crosses country borders. The internet can also be used for breaches of copyright legislation, fraud, identity theft and money laundering.

The internet has also been used to attack its own features, for example cyber-security attacks through malware or DDoS attacks. Cyber crime and cyber security have become major concerns.

> **Further details about threats to digital systems can be found in section 4.1.3 of this content area and Content area 8.**

However, carrying out surveillance can be seen as contravening the ECHR. So the question must be asked – who decides what is an acceptable reason to carry out covert surveillance?

ECPA

The ECPA is a piece of US legislation and relates to the use of a digital system relating to electronic communications. The legislation includes the Stored Communications Act (SCA) which covers the content of emails, private Facebook messages, YouTube videos and metadata, including websites visited, sender and recipient, and time/date stamps on emails.

The ECPA protects phone and electronic communications while those communications are being made, are in transit and when they are stored on computers.

However, as with the ECHR, it is possible to apply for permission to carry out covert surveillance if there is a perceived threat to national security or criminal activity. Although the ECPA is not relevant in the UK,

if a UK company has a US presence then it would be in the relevant international offices.

CAN-SPAM

The CAN-SPAM is another piece of US legislation and relates to the use of a digital system to send unsolicited commercial email.

The Act bans incorrect, deceptive or misleading subject information and lines. The Act requires that unsolicited commercial email is identified as an advertising email. As with the DPA and GDPR this Act requires that those people who receive advertising email are supplied with the means to opt-out. The Act also establishes the criteria for determining the primary purpose of a commercial email.

In addition, the Act directs the US Federal Trade Commission (FTC) to issue rules about the subject lines of emails containing sexually explicit contents. Although the CAN-SPAM is not relevant in the UK, if a UK company has a US presence then it would be in the relevant international offices.

> **Further details about the DPA and GDPR can be found in section 4.1.2 of this content area.**

Test yourself

1 What does EHCR stand for?
2 What does Article 8 of the EHCR aim to protect?
3 Identify one exception to Article 8 of the EHCR.
4 What does the ECPA protect?
5 Under the CAN-SPAM what do advertising emails need to include?

4.2 Guidelines and codes of conduct

4.2.1 The purpose and role of codes of conduct produced by professional bodies for the use of digital

There are three main professional bodies connected to digital professionals working in the digital industry. These are:
► the British Computer Society (BCS)
► the Institute of Analysts and Programmers (IAP)
► the Association for Computer Machinery (ACM).

Each of these professional bodies has developed a **code of conduct** for their members. The purpose of the codes of conduct is basically the same – to provide a set of guidelines which members agree to abide by.

Each of the professional bodies are relevant to different job roles within the digital industry. It is, however, possible to belong to more than one professional body.

- The BCS is a professional body that represents those working in IT and computer science.
- The IAP is a professional body that represents those working in the digital sector as an analyst or programmer.
- The ACM is aimed at a broad range of computing professionals.

Each of these professional bodies not only acts as a community for those working in the digital industries but also provides:

- training and a range of qualifications to enable members to update their skills
- updates in the advances in the digital industries
- a range of resources that can be used by members to enhance their own knowledge.

Each of the professional bodies has links with education including schools, colleges and universities.

Key term

Code of conduct: a document which defines rules, values, ethical principles and vision.

The **purpose** of a code of conduct is to provide members with a set of guidelines and principles which need to be followed to uphold the philosophy of the professional body. When members join a professional body, they are agreeing to follow the guidelines and principles set down in the code of conduct.

The **role** of a code of conduct is to provide guidelines for members. These guidelines provide clear details about how members should behave and conduct themselves in their professional lives.

The BCS states on their website that their code of conduct serves as a:

> 'unique and powerful endorsement of your integrity and also serves as a code of ethics for IT professionals.'

> (BCS, 2022a)

The IAP clearly states on their website that they need to ensure that their members hit the highest professional standards, are committed to their own development and abide by their duty to serve the public interest in their work.

The ACM states that the actions of computer professionals change the world, and that their members should reflect on the impacts of their work to ensure it supports the good of the public.

The codes of conduct of each of the professional bodies all include the commitments of the members to work in the public interest and to accept their professional duties. The BCS also includes the statement that members accept their professional duty, this is the:

> 'foundation of the digital professional which is built upon everyday by the competence, integrity and diversity of their members.'

> (BCS, 2022b)

The codes of conduct of each of the professional bodies also act as a benchmark when assessing the misconduct of their members. Each code of conduct includes statements of responsibility related to the IT and computing digital industry.

The main points of the code of conduct for the BCS are:

> 'You make IT for everyone.
> Show what you know, learn what you don't.
> Respect the organisation or individual you work for.
> Keep IT real, Keep IT professional, Pass IT on.'

> (BCS, 2022b)

Research

Investigate the codes of conduct for the IAP, ACM and BCS. Compare the codes of conduct for the three professional bodies. Evaluate the similarities and differences between each code of conduct. Identify any possible omissions.

Create a digital communication to present your findings to a group of students aged 17–19 years old.

4.2.2 The guidelines provided in professional codes of practice

Some professional bodies and professions related to the digital industry have a **code of practice**. A code of practice usually includes a code of conduct but also includes standards for practice. A code of practice is one part of the legislation, practice standards and employers' policies and procedures that their employees must meet. However, some organisations use the terms conduct and practice interchangeably.

The difference between a code of conduct and a code of practice

A code of conduct is a set of guidelines that sets out actions that members may or are advised to follow. A code of practice makes these guidelines mandatory.

Digital professionals and those other professionals, such as managers, who wish to comply with a code of practice must follow exactly what is set down within the code, to comply with and be covered by it.

A code of practice usually also covers applicable legislation. This may include, for example, where personal data is stored, details of the DPA or the application of the CMA to those employees working with digital devices.

Other areas covered by a code of practice may include:

▶ **professional responsibilities** including quality of work, meeting deadlines, communication, confidentiality and trust
▶ **contribution to society** including how actions have an impact on society. It should be defined as being a positive contribution to society, while actions that fail to comply will have a negative effect

▶ **safety** can cover a wide range of details. The details contained in this section of the code of practice will depend on the sector or job role that the code of practice refers to
▶ **security and privacy** include where data and information will be stored, what processes should be used to maintain the security, and how the privacy of the data and information will be maintained
▶ **innovation** which could include, for example, who has ownership of the outcomes of any tasks undertaken by employees or the confidentiality required from employees when working on, for example, new innovations in the digital industry.

Ownership of innovation is covered by the IPA in section 4.1.5 of this content area.

4.2.3 The impact of implementing guidelines from professional codes of practice on organisations and their stakeholders

Environmental

Many environmental issues have been caused by the increased **use** of digital devices. A further environmental impact is related to the increased **production** of digital devices to keep up with consumer demand. This has had an impact on legislation and professional codes of practice, which in turn has an impact on the practice of organisations and their stakeholders.

> **Activity**
>
> The BBC ran a news article in August 2019 detailing the impact of smartphones on the environment. Find the article on the BBC website and create an infographic providing details of the impact of smartphones on the environment.

In an attempt to reduce the number of digital devices being sent to landfill, the UK Government created the Waste Electrical and Electronic Equipment (WEEE) Recycling Regulations 2013. These regulations became part of the UK law on 1 January 2014. An update to the regulations came into force on 1 January 2019. The update increased the range of products to be covered by the regulations. Many consumers of digital devices appear to be unaware that if a new device is bought, the retailer now has a duty to take the old device for recycling.

One of the impacts of having to implement the WEEE regulations is that retail businesses have to provide recycling facilities for customers and arrange to get these items collected by registered waste disposal firms. This all adds to costs which may be passed on to the customer.

Staff will also need to be trained on WEEE regulations. This may be completed in-house, at a training centre or on-line. Whichever method is used, training staff will have an impact in terms of costs and time. The code of practice may also include details relating to the environmental impact within a business or organisation. These details may include such points as:

- thinking before emails are printed
- emailing documents, for example reports, rather than printing and distributing them
- using cloud-based areas to share documents with a team
- recycling and/or repairing equipment.

One of the impacts of including these points in a code of practice will be a reduction in costs of, for example, consumables including paper and ink used for printing.

> **Activity**
>
> Make notes about other points that could be included to negate the environmental impact, how these could be achieved and the impact of these points on an organisation and its stakeholders. Discuss your thoughts with the rest of your group. During your discussions, make a group list of all the points you have discussed.

Ethical

Ethics are defined as 'the moral principles, or rules, that govern a person's attitudes and behaviour'. Ethics apply not only to the way we live our day-to-day lives but also to the use of digital devices by individuals and businesses. The main areas of ethical concern relating to the use of digital devices are the privacy of stored data and cyber security.

Stored data can be easily copied (intercepted at rest) or intercepted during transit. Anyone, including an individual, charity, club, business or organisation, has a duty of care to those people whose personal data and information they have stored on a digital device. The main piece of legislation relating to how stored data should be dealt with is the Data Protection Act.

> The DPA is covered in section 4.1.2 of this content area.

The main impact of adhering to the DPA is that the business or organisation will not be prosecuted. This means that fines for non-compliance will be avoided, and the business or organisation will be perceived by customers, clients and the wider public to be trustworthy.

A code of practice may include other points related to ethical impacts. How much detail and what is included will depend on the sector the code of practice relates to. For example, a financial organisation will probably include details relating to who can access financial information and the security procedures that have to be followed by a customer. Financial details can be useful to attackers who can use the details to commit, for example, fraud or identity theft. Details about confidentiality may also be included in the code of practice.

The main ethical consideration is that of the confidentiality of data and information. Much of the data and information held by a business or organisation can be seen to be confidential, for example personal records and customer contact details. It would be unethical for this data and information to be shared or put into the public domain.

Employees should be trained on how data and information should be accessed and who the data can be shared with. It is also important that employees understand the consequences of not following the code of practice, which may include termination of a contract of employment and prosecution under the appropriate legislation. The impact of training employees is the cost and time taken to complete the training. A positive impact is that, if employees are aware of the procedures relating to the confidentiality of data, the business, organisation and employee are unlikely to breech the legislation so avoiding any prosecution.

If a business, organisation or employee is prosecuted then this can lead to a negative reputation.

When data is being collected, stored and processed the ethical implications and considerations defined in the code of practice may relate to:

▶ who is permitted to access, edit and process data, to maintain the confidentiality and integrity of the data
▶ the staff training that is to be completed before data can be accessed, including updating training on a regular basis to ensure that the data is handled ethically
▶ when updates are completed on data and the procedures, for example, for identifying the data holder
▶ consideration about new systems and adaptations to current digital systems for all end-users including those with accessibility needs. This will

ensure that all end-users are treated fairly and in line with guidelines to ensure that the protected characteristics as defined by the Equality Act.

Equality Act is covered in section 4.1.4.

One of the impacts of including and adhering to these points in a code of practice is that data will be safe meaning that the risk of breeches will be reduced which, in turn, will lead to a reduction in the risk of any legislation being broken. This will lead to a reduction in the risk of an organisation being fined as these fines can be large. Any fines will have an impact on the profitability of a business or organisation.

Activity

Make notes about other points that could be included to reduce the ethical risks to stored data, how these could be achieved and the impact of these points on an organisation and its stakeholders. Discuss your thoughts with the rest of your group. During your discussions, make a group list of all the points you have discussed.

With the increasing reliance on the use of technology and digital devices in everyday life, a further ethical and moral dilemma has emerged. There is an assumption that everyone has access to digital devices and the internet, and that users can use this technology as intended. This is known as the 'digital divide'.

In the context of ethics, a code of practice may also include points that relate to the digital divide. The main factors in the digital divide are:

▶ Wealth: this relates to the amount of disposable income a person has. Disposable income is the income left after financial commitments including utility bills, loans and food have been paid for.
▶ Location: this is where the end-user is located. This can be defined as local, national or international. The location can have an impact on the connection speed and mobile data accessibility.
▶ Specific needs: these may be due to visual, auditory or physical disabilities. These can limit accessibility to digital devices and the WWW. Article 9 of the United Nations (UN) Convention on the Rights of Persons with Disabilities (UNCRPD) defines the specific rights of people with disabilities to access digital devices.

▶ Ability to use new technology: this can relate to those users with specific needs but also to those who lack confidence or the ability to keep up to date with the fast pace of updates and new digital devices.

> You will learn more about accessibility later in this content area in section 4.2.4.

You will learn more about accessibility later in this content area in section 4.2.4.

Research

Investigate the UNCRPD. Create an information sheet detailing the different features that can be utilised to increase accessibility for those users with a disability.

Again, what is included may depend on the sector the code of practice applies to. For example, telecommunication companies may have a point in the code of practice that vulnerable customers, such as those with a disability, will be a priority when communication faults, e.g. problems with broadband, are being resolved. . Employees will be made aware, usually provided in the code of practice, of the factors that define a vulnerable customer.

> The Equality Act is covered in section 4.1.4

The Equality Act is covered in section 4.1.4

Points in the code of practice may include:
▶ Being aware of situations relating to vulnerable customers and who to escalate these to.
▶ Designing and producing devices that are accessible to all end-users.
▶ Ensuring that devices are able to backwards use a range of connectivity protocols
▶ One of the impacts that may result from consideration of the different factors of the digital divide include enhanced reputation for being inclusive. Another impact may include recognition by a range of organisations who address issues relating to accessibility issues for those customers who have specific needs.

Activity

Make notes about other points that could be included that relate to the digital divide, how these could be achieved and the impact of these points on an organisation and its stakeholders. Discuss your thoughts with the rest of your group. During your discussions make a group list of all the points you have discussed.

Activity

In a group, revisit the codes of conduct from the BCS, IAP and ACM. Link the contents of the codes of conduct with the ethical and moral issues raised in this section. Present your findings to the rest of your class. Discuss the similarities and differences identified by each group.

Legal

A code of practice may include points that relate to specific legislation. As before, what is included will depend on the sector the code of practice applies to. For example, a sector that stores credit card data will need to ensure they comply with the DPA and the CMA. However, most organisations that store data, such as payroll and personnel data, will need to ensure it is secure to comply with the DPA.

Details may also be provided that relate to the safe use of display screen equipment (DSE) and other work-based activities such as working at height. Any changes to the health and safety legislation relating to the use of DSE will need to be considered, updated and the changes implemented. There may be an increased cost implication of implementing any changes to the health and safety. This could include the purchase of new equipment, not just that related to digital systems but also to equipment used when working at height. Employees will also need to be trained to use new digital systems or other equipment.

While this may be seen as an added expense to a business or organisation, if new equipment is not purchased or employees are not trained then this could lead to an accident resulting in possible injury to employees. As the health and safety legislation was not followed by a business or organisation then they would be in breach of this legislation, which could, in turn, lead to fines from the HSE.

Points in the code of practice may include:
▶ compliance with the CIA triangle which also ensures compliance with the DPA and CMA
▶ security processes and procedures, physical and logical, that are used to mitigate against security threats. The DPA (2018) includes the principle 'Handled in a way that ensures appropriate security, including protection against unlawful or unauthorised processing, access, loss, destruction or damage
▶ keeping access details secret to reduce the risks of a data breach and increase compliance with the DPA / GDPR

- staff training to ensure knowledge related to compliance with relevant legislation
- creation and updating of policies and procedures relating to legislation. For example, if the DPA is updated, changes to meet the update should be incorporated into procedures detailed in policies and procedures to ensure continued compliance.

One of the impacts that may result from including and implementing these points will be an increase in the security of stored data leading to a reduction in security breaches. These security breaches could be intentional or caused by human error. By including these, staff will be aware of the processes and procedures that need to be followed which should, in turn, make them more aware of how data should be handled.

> DPA / GDPR is covered in section 4.1.2.

> CMA is covered in section 4.1.3.

> International legislation related to data use is covered in section 4.1.7.

Activity

Make notes about other points that could be included that relate to legislation, how these could be achieved and the impact of these points on an organisation and its stakeholders. Discuss your thoughts with the rest of your group. During your discussions, make a group list of all the points you have discussed.

Test yourself

1 What is the main cause of environmental issues relating to digital devices?
2 What is meant by the term 'ethics'?
3 What is meant by the phrase the 'digital divide'?
4 What are the four main factors related to the digital divide?
5 What does Article 9 of the United Nations Convention on the Rights of Persons with Disabilities (UNCRPD) cover?

4.2.4 How guidelines and agreed standards ensure the accessibility and quality of IT systems

The UNCRPD provides some details relating to the accessibility of information systems and digital devices. There are four main organisations that aim to increase this accessibility and provide other standards and guidelines focused on a range of issues relating to the use of digital devices, data and information and cyber security.

ISO

The International Standards Organisation is an independent, international, non-governmental organisation. The ISO currently has 167 national standards bodies as members. It was formed in 1946 in London and has grown to become one of the most influential organisations in the creation and application of international standards across a range of sectors. The aim of the ISO is to:

> 'bring together experts to share knowledge and develop voluntary, consensus-based, market-relevant international standards that support innovation and provide solutions to a range of global challenges.'

(ISO, 2022)

Research

Visit the ISO website (www.iso.org). Look at the timeline about its creation and development. Make a list of all the sectors that are affected by the work of the ISO.

Cyber security is high on the list of priorities for governments around the world because of an increased level of cyber threats. These threats have come about from the increased use of digital devices and the internet which has made the world appear smaller and more interconnected.

The ISO has developed the '27000 category' of standards. The focus of these standards is on the safety of information assets. These assets include personal and organisational data as well as the confidential, national security and government data stored in the cloud.

These standards also assist organisations to manage the security of their assets such as intellectual property, and financial and employee data. This also includes the cloud providers who hold information in trust for third parties.

There are more than 12 standards in the 27000 suite. Some of these and their application areas are shown in Table 4.9.

Standard	Application area
ISO/IEC 27000:2018	Provides an overview of information security management systems (ISMS). Terms and definitions commonly used in the ISMS family of standards are also included. This relates to all types and sizes of organisation (e.g. commercial enterprises, government agencies, not-for-profit organisations).
ISO/IEC 27001	Stipulates and defines the specifications for the implementation of an ISMS.
ISO/IEC 27002:2013	Provides guidelines for organisational information security standards and information security management practices including the selection, implementation and management of controls taking into consideration the organisation's information security risk environment(s).
ISO/IEC 27004:2016	Provides guidelines intended to assist organisations in evaluating the information security performance and the effectiveness of an ISMS in order to fulfil the requirements of ISO/IEC 27001:2013, 9.1.

▲ Table 4.9 Some of the standards in the 27000 suite and their application areas

Many confuse the internet with the World Wide Web – the Web.

The internet is defined as:

> 'the global network comprising a loose confederation of interconnected networks using standardized communication protocols, which facilitates various information and communication systems such as the World Wide Web and email.'

> (OED Online, 2022)

While the World Wide Web is defined as, according to W3C:

> 'An information space in which the items of interest, referred to as resources, are identified by global identifiers called Uniform Resource Identifiers (URI).'

> (W3C Architecture of the World Wide Web, 2004)

Activity

Create an information sheet for a new member of staff in an IT support department detailing the ISO standards relating to IT systems and cyber security. You should provide details of each standard and how they can be complied with.

Important point

There is a difference between the internet and the World Wide Web. In this book the term 'the internet' has been used as most people use the internet as a blanket term to include both.
It is, however, important that you understand the difference between the internet and the World Wide Web.

W3C

The World Wide Web Consortium (W3C) is an international community where organisations, staff and the public work together to develop web standards. W3C is headed by Tim Berners-Lee and Jeffrey Jaffe.

The mission statement of W3C is:

> 'To lead the Web to its full potential.'

> (W3C, 2022)

Tim Berners-Lee is credited with inventing the World Wide Web. In 1989 he wrote a proposal which then led to him writing a browser, server and webpage. He also then went on to create the first specifications for **Uniform Resource Locators (URLs)**, **Hyper Text Transfer Protocol (HTTP)** and **Hyper Text Markup Language (HTML)**.

URL: Uniform Resource Locator. The address of a website/page. The URL also provides details of how to reach a website or webpage, for example http.

HTTP: Hyper Text Transfer Protocol. The communication protocol that is used when connecting to web servers. The servers can be on the internet or an intranet.

HTML: Hyper Text Markup Language. The markup language that is used when creating documents to be used on a website or webpage. HTML can be used to denote different elements such as font style/size and colours.

The main focus of W3C is to develop protocols and guidelines that ensure long-term growth for the Web. W3C's standards define key parts of what makes the World Wide Web work. These standards cover:

▶ web design and application
▶ web of devices
▶ web architecture
▶ semantic web
▶ extensible markup language (XML) technology
▶ web of services
▶ browsers and authoring tools.

Activity

In a small group, select one of the W3C standards. Investigate the standard you have selected. Present your findings to the rest of your teaching group.

Web accessibility is one of the main focuses of the W3C. Part of the work completed by the W3C is the development of the Web Content Accessibility Guidelines (WCAG).

Web Content Accessibility Guidelines (WCAG) 1.0 and 2.0

The WCAG created by W3C, aim to define a standard for web content accessibility that meets the needs of individuals, organisations and governments worldwide. The WCAG complements those guidelines defined by the UNCRPD.

WCAG created V1.0 of the guidelines in 1999. At that time, the use of the internet and World Wide Web as a tool was not as prevalent as it is today: just 4.1% of the world's population, 248 million people, were users, with email becoming a commonly used method of communication.

WCAG V1.0 set down 14 guidelines that covered the issues that may be faced by those people who have a disability. WCAG V1.0 attempted to make web developers realise that not all users had the same needs. These needs include disability, software versions and technology different to that used to develop the content.

The 14 guidelines from WCAG V1.0 are:

1 Provide equivalent alternatives to auditory and visual content.
2 Don't rely on colour alone.
3 Use mark up and style sheets and do so properly.
4 Clarify natural language usage.
5 Create tables that transform gracefully.
6 Ensure that pages featuring new technologies transform gracefully.
7 Ensure user control of time-sensitive content changes.
8 Ensure direct accessibility of embedded UIs.
9 Design for device-independence.
10 Use interim solutions.
11 Use W3C technologies and guidelines.
12 Provide context and orientation information.
13 Provide clear navigation mechanisms.
14 Ensure that documents are clear and simple.

Activity

Watch the video entitled Introduction to Web Accessibility and W3C standards: www.w3.org/WAI/videos/standards-and-benefits.

Answer these questions:
1 What features are defined for accessibility?
2 What situations are detailed?
3 How can accessibility features be included? Provide some examples.
4 How does each feature work?
5 What are the three sets of guidelines created by W3C?
6 What features does the video include, or that you can access, to conform to the accessibility guidelines?

As a group, discuss your findings.

Although these guidelines were developed in 1999 many of them can still be applied when creating web content today.

Activity

Research the 14 guidelines of WCAG V1.0. Assess your centre's or employer's website against each of the 14 guidelines. Create a communication to be presented to the website owner, detailing compliance to the 14 guidelines. Make suggestions as to how the website could be improved.

Web content, according to the WCAG, is the information on a webpage or web-based app including:

▶ natural information such as text, images and sounds
▶ code or mark-up that defines, for example, structure and presentation.

As the number of hosted websites has increased, so the WCAG have been updated. WCAG 2.2, introduced in 2021, subsumes the contents of 2.0 and 2.1. WCAG 2.0 is approved as an ISO standard: ISO/IEC 40500:2012.

The WCAG has four guiding principles. These are:

▶ perceivable
▶ operable
▶ understandable
▶ robust.

Table 4.10 shows each principle, a definition and a meaning.

If any of these are not true, then users with disabilities will not be able to access and use websites.

Activity

Compare the 14 principles defined in V1.0 with those principles and guidelines defined in the current WCAG. Make notes about your findings.

Contained in WCAG 2.1, there are 13 guidelines split between the four principles. Each guideline has success criteria. The guidelines demonstrate how web content can increase accessibility to those with a disability. However, many of the principles and guidelines can also help in making web content more accessible to all users.

Activity

Complete the table to show the principles and the associated guidelines.

Principle	Guideline(s)
Perceivable	
Operable	
Understandable	
Robust	

WCAG recommends testing any new web content, which is also good practice. The success criteria have been written to be software and device non-specific. Some of the success criteria can be tested using software testing applications. It is also recommended that usability testing, with a range of users, is carried out.

Principle	Definition	Meaning
Perceivable	Information and UI components must be presentable to users in ways they can perceive.	The content and components cannot be invisible to all of the user's senses.
Operable	UI components and navigation must be operable.	The interface cannot require interaction that a user cannot perform.
Understandable	Information and the operation of the UI must be understandable.	The content and operation of the UI cannot be beyond the user's understanding.
Robust	Content must be robust enough that it can be interpreted reliably by a wide variety of users, including those using assistive technologies.	As technologies evolve, the content should remain accessible.

▲ Table 4.10 The four guiding principles of the WCAG

Research

Access the current WCAG. Using the same website that you applied WCAG V1.0 to, apply the current WCAG. Does the website comply with the updated WCAG? What could be done to make the website more compliant? Discuss your findings with the rest of your teaching group.

Internet Engineering Task Force (IETF)

The IETF is an internet standards organisation. The members are international and are mainly working as network designers, operators, vendors and researchers.

The mission of the IETF is:

'to make the internet work better by producing high quality, relevant technical documents that influence the way people design, use, and manage the internet.'

(IETF, 2004)

The main focus of the IETF is the 'evolution of the internet architecture and the smooth operation of the internet'. Looking back in this section, you have learned that the internet is defined as the hardware on which the World Wide Web presents the information.

The IETF have five guiding principles. These are:
- open process
- protocol ownership
- rough consensus and running code
- technical competence
- volunteer core.

The main areas of interest of the IETF change over time as technological developments change. This means that these areas keep up to date with the technological developments. The main areas of interest for the IETF in 2021 were:
- augmented network management
- the internet of things (IoT)
- new transport technology
- security and privacy.

Most of the work of the IETF is carried out by remote working practices. These include virtual conferences, mail lists and forums. The IEFT issue **RFCs**. These contain technical and organisational notes about the internet. The IETF have a library of RFCs which can be searched using keywords. Once an RFC has been published in the library it never changes. There are,

however, occasions where errors have been identified. If errors are identified, then an erratum is uploaded to the library. Errors and the associated erratum can be categorised into:
- technical – errors identified in the technical content
- editorial – errors identified in spelling, grammar, punctuation, or syntax errors that do not affect the technical meaning.

Key term

RFC: Request for Comments.

All of the organisations covered in this section have the same basic aim. That is to make sure that IT systems are accessible and of a high quality. It is important that every professional working on projects connected to IT systems uses the same basic protocols and standards. By doing this, the functionality of any IT system will be consistent and accessible for all users.

To turn this on its head – what would happen if every professional used a different set of standards and protocols?

Test yourself

1 Identify and describe two principles set down by the WCAG.
2 Identify three standards defined by the W3C.
3 According to the WCAG, what is web content?
4 What are the four principles of WCAG 2.1?
5 What is the main focus of the IETF?
6 What are the two categories of errors identified by the IETF?

4.2.5 The role and implications of an acceptable use policy (AUP) within an organisation

An AUP is an agreement between employers and employees. An AUP sets down the rules that a user must agree to for access to a network, email facilities or the internet. The AUP may also set down guidelines as to how the IT systems should be used. Many businesses, organisations, schools and colleges require a signed AUP before log-in details are provided.

One reason an AUP may be created is to set down points that staff, students, employees and employers can adhere to. The AUP can also be used where the rules included in the AUP have been breached. These breaches may lead to disciplinary action or, in the most severe cases, dismissal.

However, the AUP can also protect the employer against any actions taken by an employee which results in legal action, for example if an employee is sending nuisance emails from their work email address.

Some of the areas that may be covered in an AUP include:

▶ what monitoring may be carried out by the employer in the workplace
▶ activities that are not allowed, for example the use of personal email
▶ non-acceptable use of, for example, the internet, email, websites and social media
▶ procedures for the use of systems and removable storage devices
▶ taking and using business digital devices off-site.

The monitoring of the workplace is covered in section 4.1.6 of this content area.

Different areas of an AUP will include specific points relevant to that area.

For example, the area relating to non-acceptable uses of the internet and email may include:

▶ using the internet or email for the purposes of harassment or abuse
▶ using profanity, obscenities or derogatory remarks in communications

▶ accessing, downloading, sending or receiving any data (including images), which is considered to be offensive in any way, including sexually explicit, discriminatory, defamatory or libellous material
▶ using the internet or email to make personal gains or conduct a personal business
▶ using the internet or email to gamble.

The area relating to removable storage devices may include:

▶ It is not permissible to connect any non-authorised device to the network or IT systems.
▶ The storage of data on any non-authorised equipment is not permitted.

The area relating to taking and using business digital devices off-site may include:

▶ Equipment and media taken off-site must not be left unattended in public places and not left in sight in a car.
▶ Laptops must be carried as hand luggage when travelling.

An AUP must be tailored to the needs and requirements of the, for example, business which is creating it. There are many templates and sample AUPs provided on websites. These can be used as a starting point, but the specific needs of the business must be included.

Skills practice

A small start-up company delivers pet food for its customers. Customers order pet food by phoning one of the customer support team.

The pet food is stored in a warehouse. Each order is picked and packed in boxes by the warehouse employees. The boxes are delivered by a courier company.

Business is improving with more customers placing orders. The company is employing new employees to work in their offices and warehouse.

The company is considering creating a website where customers will be able to register and order their pet food. Customers will need to provide their name, address, phone number and an email address when registering.

The customer details will be stored by the company. The company wants to send emails to the email address provided on registration to tell their customers about new products and services.

You have been asked to:
▶ Provide details of threats to the security of the stored customer and company data and information.
▶ Provide details of the legislation relating to the storing and use of customer data and the actions needed to comply with this.
▶ Explain the importance of following accessibility standards and guidelines relating to the creation of the website.
▶ Provide details, including justifications, about the accessibility features that should be included on the website.

Assessment practice

1 Explain, using examples, the importance of the Health and Safety at Work Act when working with DSE.

2 Discuss the rights of a data subject under the DPA.

3 Identify and describe two different types of social engineering.

4 Identify and describe three different types of discrimination that can be applied to the protected characteristic of disability.

5 Discuss why an employer may want to monitor electronic communications in the workplace.

6 Discuss why it is important that the members of a digital professional body agree to the code of conduct.

7 A company stores details of customers. Explain why their code of conduct should include details relating to IT legislation.

8 Explain how the move to online retail may impact on the older generation.

9 Discuss why it is important that websites should not be device dependent.

10 Identify two areas that could be included in an AUP. For each, explain why it is important these are included.

Content area 5: Business context

There are many factors that affect every business, and these include internal and external factors. In addition, these factors can have both a positive and a negative impact on the business. These factors are the business environment. Factors include human resources, the quality of management, finance, marketing, logistics, research, design and development, as well as the political climate and environmental concerns. The range of factors that can affect the business environment are vast and within this unit you will learn about its common features and whether the factors are internal or external factors.

There are different types of organisations in a wide variety of sectors. All these businesses need, and must be able to use, digital technologies. Implementing and extending the use of digital technology requires careful strategic and project planning. This means considering not only the hardware, software and storage requirements, but also the needs of the end user, the business and its stakeholders, and social, political and legal factors.

Learning outcomes

In this content area you will learn about:
5.1   The business environment
5.2   Digital value to business
5.3   Technical change management
5.4   Risks in a business context

5.1 The business environment

5.1.1 The purpose of different types of organisations in a range of sectors

Businesses provide products, a service, or even a product and a service. It is important to understand the difference between a product and a service.

Products

Products fill the need and want of the customer.

Products come in many forms. Clothes can be purchased in different styles, colours and sizes, cars can be purchased from a wide range of manufacturers, models, colours, sizes and styles. A product is often a physical thing, but it can also be something with no material presence, for example software or a package of digital teaching resources. The quality of products can be compared, and a product can be more easily returned if it is faulty, not suitable, or not as advertised. Products do not have to be used straight away. They are available for when they are needed and wanted.

Services

While services can also meet a need or want, they are more often about building relationships and trust. When you visit a doctor or a dentist, you may not necessarily leave with something tangible, such as a prescription. You may go to the dentist for a regular check-up on your teeth and gums. All being well, you leave without taking a physical object away with you.

It is much harder to compare the quality of services. If you have applied for a passport, you cannot make a judgement on the quality of the service until you have received your passport without any issues or delays. It is difficult to return a service because a service is used as it is being offered. People can claim refunds for poorly delivered services, but it is not as easy as returning a product.

One important thing to remember about a service, once offered and not used, it is lost forever. If you book accommodation for a holiday and then do not travel to the destination and stay at the hotel, that time, date and so on is lost forever.

It is also important to consider that as technology provides more and more products, the service element (e.g. customer service, online sales, instruction manuals, online support services) also increases. Therefore, products can also include elements of services.

Test yourself

1 A person purchased a cooker and has lost the instruction manual. They go online to the manufacturer's website to download a copy of the instruction manual. Is this a product, a service or both?
2 You go to the supermarket to buy food. Is this a product, a service or both?
3 You are buying a new car and the showroom offers you a deal on the insurance. You decide to buy the car you want and purchase the insurance. Is this a product, a service or both?
4 You work for an IT company which installs networks. They sell service level agreements to their customers should problems arise. Does the company offer a product, a service or both?
5 You have attended an eye test at the optician. The optician confirmed that you do not have any issues with your sight or your eyes. Have you purchased a product, a service or both?

Business sectors

There are four main categories of business sectors: primary, secondary, tertiary and public. We are going to look at each of these categories in more detail.

Primary

The primary sector is concerned with the extraction of raw materials such as mining, oil and gas, and fishing industries, as well as the growing of goods such as farmers growing crops of wheat, and cocoa and coffee plantations.

▲ Figure 5.1 Oil is one of the raw materials extracted by the primary sector

Secondary

The secondary sector's function is related to manufacturing. This invariably involves using the raw materials purchased from the primary sector and manufacturing them into new products. The building, food processing and car manufacturing industries are all examples of businesses that operate in the secondary sector.

▲ Figure 5.2 Car manufacturing is an example of a secondary sector business

Tertiary

The tertiary sector focuses on the provision of services. Banks, supermarkets, cinemas and hairdressers are in the tertiary sector. Supermarkets sell the food produced and/or packaged from the secondary sector. Hairdressers use products manufactured in the secondary sector. Banks offer services which include securely holding a person's money and selling mortgages and loans. Cinemas show the films produced by the film industry. In addition, industries providing information services such as computing, information and communication technologies (ICT) consultancy and research all come under the tertiary sector. The tertiary sector makes up the largest part of the UK's economy.

Public

The public sector includes any organisation that is owned and operated by government agencies. Examples of public sector organisations include the local council, His Majesty's Revenue and Customs (HMRC), the passport office, libraries and schools. Although it can be argued that these organisations provide a service, they are not classified in the same category as the tertiary sector.

Test yourself

Complete the table by identifying the sector that each business comes under.

Business	Sector
1 Police force	
2 Oil rig in the North Sea	
3 Radio station	
4 Dairy farm	
5 Chocolate factory	

5.1.2 The key areas of organisations and how IT is used to support them

Organisations implement many processes to function effectively and smoothly. Each of the processes is carried out by a particular functional area. But each functional area does not function in isolation. There must be collaborative working between the functional areas and digital technology plays a big part to support this. Functional areas may not always be in the same building; they could be in a different building and even different locations.

Human resources

The main function of human resources (HR) is to recruit people to undertake a variety of roles within the organisation. To do this effectively, they must ensure that people have the appropriate qualifications, experience, knowledge and skills for the position they are applying for.

HR is also responsible for the training of employees, employee relations, understanding and keeping up to date with current workforce legislation and regulation, setting pay scales, benefits and compensation. The HR functional area is one of the most critical within any organisation as it deals with people. The workforce is important to any organisation.

Use of IT

HR uses technology in many ways including communicating with other functional areas, contacting employees, analysing employee performance and storing information. When used effectively, IT can improve the efficiency of the HR functional area. If used inefficiently, it can disrupt the way the area functions. Digital technology makes it easier to gather data on employees to get an overall picture of the functionality and performance of the business.

Many HR departments post job opportunities online and ask potential applicants to apply through an online tracking system. While this provides the department with more time to carry out other functions, the system is not always effective at identifying the potential in applicants. Forms are standardised and a badly designed system can be cumbersome to navigate and deter potential applicants from applying.

Communication is now easier with the use of email, text and messaging apps. HR staff can maintain contact with other functional areas within the organisation. If a manager wants to share information with the logistics functional area about a new employee, they can send an email to the relevant team members. It is important that there is not an over-reliance on technology as lengthy emails are counter-productive and may be more appropriately delivered in a face-to-face meeting or using video conferencing.

There are software programs available that can support the analysis of employee performance, providing results such as whether an employee finishes their work on time and to standard. How does their performance compare with the previous year/month? Has it increased or decreased, and by how much?

It is important that HR does not obtain too much data on employees, or they could feel that their privacy is being impacted, for example using cameras to record what employees are doing. Being constantly monitored can alienate employees and it is therefore important that HR strikes a good balance between how much data *can* be obtained and how much *should* be obtained. There is always the risk of having too much data and therefore it becomes time-consuming to select relevant information. In addition, it can lead to information being misread and/or misinterpreted, or assumptions being made.

Information which is sensitive and confidential (such as employee personal records) also requires good security measures to be implemented. There should also be relevant policies and procedures in place stipulating who can have access to the data.

Research, design and development

This particular functional area plays a major role in the product life cycle of an organisation. The research, design and development functional area is separate from the other functional areas, but the functions are often related and require collaborative working. An understanding of the functions of the research, design and development functional area will help you to understand how digital technology can be used.

New product research

Before any new product is produced, a thorough study is conducted to support the project. This includes establishing the specifications of the product, the costs involved and production timescales. The research can also include an evaluation of the need for the product, in other words, is it a product that customers/ stakeholders will want to use?

New product development

Once the research has been carried out and the outcome of the research is positive, then the new product is designed and developed based on the identified specification and needs. The product must meet the product guidelines and any regulatory requirements.

Existing products

There may be a requirement for existing products to be updated. This will also come under this functional area. Research will be carried out into the feasibility of updating an existing product, or whether it would be more appropriate to design a new product, or stop production of the existing product altogether. There are occasions when the functional area will be asked to resolve a problem with an existing product or to find a solution if the production process has to change.

Quality control

This functional area can also be responsible for the quality checks of the products created by the organisation. This is because the department will have detailed knowledge of the specifications and functionality of the product being checked. This enables the team involved to ensure that the products meet the required standards. If an organisation has a separate quality control department, they will collaborate with the research, design and development functional area.

Staying competitive

Another role of this functional area is to help the organisation to stay competitive. They research and analyse their competitors' products as well as the new and emerging trends within their industry.

Use of IT

This functional area will use digital technology for communication and storage similar to HR, and they will use the internet for research on competitors.

IT might also be used for:
▶ Product design: product developers will use CAD (computer-aided design) packages to develop the designs as 2D drawings or 3D models with precise measurements.

▶ Presentation of design ideas: as well as the standard presentation software, product developers can use software to develop high-fidelity prototypes to present ideas to clients or internal stakeholders. This provides an opportunity for more productive and detailed feedback so that adaptations can be made and the full product can be developed.

▶ Design documentation: this can be produced from the CAD programs but other packages can also be used to produce digital mood boards, digital storyboards, flow charts, as well as many other forms of design documentation. This can facilitate discussions about ideas, as well as helping to manage ongoing projects.

As with HR, security of the storage of files is very important and relevant policies and procedures must be developed.

> ### Research
>
> Research the different ways that a research design and development department within the car industry would use different forms of IT and associated software. The research should include:
> ▶ the types of software used
> ▶ what they are used for.

Logistics

Logistics is the planning framework for material management, information and service. Logistics can include information that is complex, which means that communication and control systems are important to the everyday effective running of the business.

People say that logistics is having the right type of product or service at the right place, at the right time, for the right price and in the right condition. Consider Amazon and the logistics involved in distributing the vast range of different products purchased by their customers, as well as procuring additional or the same products for the warehouses to store. Amazon's functional area of logistics is critical to its success.

Logistics involves several activities which include:
▶ purchasing
▶ planning
▶ co-ordination
▶ transportation
▶ warehousing
▶ distribution.

People within the logistics team must have the relevant knowledge, skills and understanding to carry out:
▶ planning
▶ controlling
▶ directing
▶ co-ordination
▶ forecasting
▶ warehousing
▶ transportation
▶ facility location
▶ inventory management.

Use of IT

Some of the uses of digital technology are similar to the other functional areas. In addition, there are other aspects of digital technology that support the functions of logistics. Here are some more examples of technology used by logistics.

▶ **Automation** – the use of data-driven software improves the operational efficiency in machines such as package labelling and streamlining the sorting systems in the warehouse.

▶ **Robotics** – robots are designed to perform several tasks at the same time. They are used in e-commerce operations to improve efficiency and speed to meet the increasing demands for online sales. Amazon purchased Kiva robots to complete 'one-click' orders in less than 15 minutes. This task would take a human around 60 to 75 minutes to complete. This reduced operating costs by 20%.

▶ **Wearable technology** – wearable technology is also becoming standard within logistics. DHL, for example, used Google Glass within one of their warehouses to see if it would increase the speed and reduce human error during the order picking process. The results of the trial showed an improved efficiency of 25% as well as increasing employee satisfaction.

▶ **Drones** – drones are becoming popular due to their ability to deliver products quickly and, in particular, to dense and congested urban areas. UPS partnered with Zipline to start up a medical drone delivery service to deliver medical supplies in Africa.

▶ **Self-driving vehicles** – logistic organisations have started adopting self-driving vehicle technologies in storerooms and shipping yards. These include autonomous forklifts and small driverless plant trucks.

▶ **Cloud computing** – cloud computing applications provide opportunities for greater efficiency and flexibility options. The use of smart processes can cut transaction costs.

▶ **Internet of everything** – the use of temperature and humidity sensors can monitor the supply-chain with respect to quality control, for example detecting when the packaging has been tampered with.

Marketing

The marketing functional area is responsible for the promotional activities and advertising of the organisation. The most effective way to communicate with customers so that they know about the products and services offered is through a range of promotional activities and advertising. The functional area of marketing is responsible for promoting the business to generate sales and help the business expand. The functions include the creation of marketing strategies, planning promotional campaigns and monitoring the activities of the organisation's competitors.

Use of IT

While organisations still use traditional methods to market their products and services, for example advertising in newspapers and magazines or on the television, and placing signs in stores, many of them are now implementing digital technology.

They now use advertising emails, social media posts or their own website which customers can find using a search engine. Technology has clearly changed the way that organisations market their products, services and even employment opportunities.

Mobile marketing is becoming more popular with organisations communicating with customers on their mobile phones through the use of text messaging and applications. Text messages can be used to send discount codes and promotional coupons.

Technology is used a lot in stores. The use of digital signs captures the attention of customers and they are used to market specific products. This is particularly useful if a business wants to introduce new products on a regular basis. Data from point of sale (POS) systems will provide a business with information on sales patterns. This can be used to analyse customer purchasing patterns and preferences, before developing marketing plans to meet those needs and maximise sales opportunities. A business, for example, might discover that customers tend to buy more of a certain product at particular times of the year. The business can then target their marketing activities to those products and the relevant customer segments.

Finance

The two main responsibilities of this functional area are **accounts receivable** and **accounts payable**. The finance functional area is also responsible for the payroll – making sure that the employees are paid accurately and on time.

Another function involves the planning to obtain **capital** finance and managing of the organisation's fund. Within the finance functional area, finance managers plan for short-term and long-term financial capital needs and analyse any impact that the borrowing of money can have on the organisation. This functional area produces reports and financial statements that can be used for budgeting, forecasting and other decision-making processes.

Use of IT

The automation of core processes and transactions frees up the finance team so that they can focus on strategic business performance and at the same time improve their services to other key functional areas using the digital reporting tools that are available.

The effective use of financial software solutions ensures that up-to-date information is readily available to support the business when making strategic decisions. Technology also enables the finance functional area to assess the current situation of how the business is performing against the set budget. This provides an overview of the financial health of the organisation as well as a detailed analysis of any business-critical areas posing a risk to its viability.

Key terms

Accounts receivable: money that is coming into the organisation.

Accounts payable: money that is paid out by the organisation.

Capital: the money or assets owned by a person or organisation and used as investments to make more money.

Technology increases the speed of critical financial business tasks such as the production of:

▶ quarterly financial reports
▶ profit and loss statements
▶ balance sheets
▶ payroll runs (weekly, monthly or both).

Finance software is often hosted in the cloud. This means that the finance team can access information that they require wherever they are. The latest technology can also gather and consolidate data and information from other systems within the organisation wherever they are situated. The benefits in timesaving alone are enormous. The use of technology by the finance functional area has resulted in greater collaboration across functional areas regardless of their location.

Management

There are four key activities relating to good management within an organisation. These are:

▶ **Planning:**
  • setting objectives and stating the mission of the organisation
  • examining alternatives
  • determining resources needed
  • creating strategies (strategic planning) to meet objectives.
▶ **Leading:**
  • leading and motivating employees so that organisational goals can be achieved
  • communicating with employees
  • resolving conflicts
  • managing change.
▶ **Organising:**
  • creating the structure of the organisation
  • setting policies and procedures
  • designating roles and responsibilities
  • designing jobs and specifying the tasks involved
  • co-ordinating work activities
  • allocating resources.
▶ **Controlling:**
  • measuring performance
  • comparing performance to standards
  • taking action to improve performance.

Use of IT

Organisations and their employees rely on digital technology to help create efficient business practices. It is important that the management of the organisation learn new technologies and evolve with change. This can include communication via email or through VoIP.

Digital technology provides numerous effective tools and applications for tasks such as managing and organising employee workloads, providing online staff training resources, communication boards and FAQs. This results in lower downtime and increased productivity.

Online Analytical Processing Systems (OLAP) support the analysis of vast amounts of data to provide information on market trends, consumer behaviour, demographics, sales, competitors, design and development, production, distribution, stock levels and costs. The scope is vast, and data can be gathered and processed quickly to provide meaningful information.

> ### Test yourself
>
> 1 Describe two responsibilities of the finance functional area.
> 2 Identify the functional area that sets the goals and mission of an organisation.
> 3 Explain how the marketing functional area uses digital technology to support the activities it carries out.
> 4 One role of the HR functional area is to employ staff. Identify two other roles it carries out and explain how it can use IT to support these roles.
> 5 Discuss the role of the research, design and planning functional area and how it can use digital technology.

5.1.3 How digital supports the business needs of organisations

Use of digital to enable automated stock/ inventory control

In section 5.1.2 you learned about the logistics functional area. In this section we are going to look at one particular role within this functional area in more detail. But first of all, it is important to understand the difference between **stock** and **inventory** control. Stock items are the items sold to customers, while inventory includes the items for sale, as well as the materials and equipment required to manufacture them. There are four main categories of inventory:

▶ raw materials
▶ products that are in the process of being made
▶ **maintenance, repair and operations** (MRO) supplies
▶ finished products (items).

How software is used

As the demand for stock increases and stock levels increase, it is important for any organisation to monitor its stock levels. Customers will search for other suppliers if they cannot purchase what they want because an item is 'out of stock'. The same applies to inventory control. As demand increases for the products manufactured and sold, it is important for an organisation to review the stock level of raw materials and MRO supplies needed for production to continue.

Digital technology and the use of good software can automate the inventory and stock control levels. There are several features of automation software that can be used by an organisation:

▶ tracking levels of stock and inventory using QR codes and barcodes in real time and across locations

▶ label production through the generation of QR codes or barcodes

▶ compatibility with other software programs such as spreadsheets so that analysis of sales, production and so on can take place

▶ producing reports which can be saved, shared and printed

▶ multi-user access (customers, employees, functional areas) so that relevant data can be shared

▶ automatic production of quotes and/or invoices which can be shared with finance software.

How hardware is used

In order to use software to carry out activities, there must be hardware to run it on. This will include servers (or even the cloud), computer terminals, networks and printers. In addition, QR code and/or barcode readers are required to scan items for tracking stock levels. Some organisations also use the camera on a mobile device.

> Other examples of hardware that can be used can be found in section 7.1.2.

How processes are carried out

Figure 5.4 shows the various stock and inventory control processes for a manufacturing company and how different parts of the system communicate with each other. Similar processes are carried out for different types of businesses.

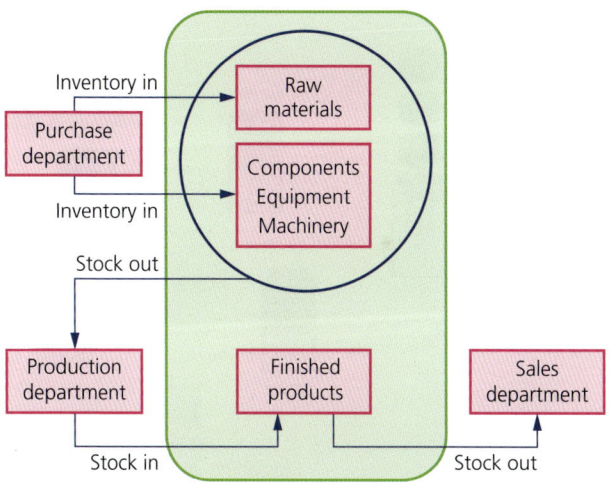

▲ Figure 5.4 The process of stock/inventory control used by a manufacturing company

Traditional and cloud-based technologies and services for communication and collaboration

The cloud is rapidly overtaking traditional in-house style systems. This is due to its reliability, scalability and cost-effectiveness as a digital IT solution. This does not mean that all businesses use the cloud instead of the more traditional systems. Businesses that have built their own data centres and IT infrastructure, which they know are robust, still rely on the traditional infrastructure for managerial and security reasons.

Selecting the correct digital model for the business is always extremely important for any organisation. Every organisation in every sector must have safe and secure storage for its data. Applications and data must be readily accessible from any location while keeping the running costs low.

Traditional technologies

Traditional technologies consist of hardware connected to a network via a server (which can be remote). The server gives the employees access to the business's stored data and applications (based on security permissions of course). Businesses using the traditional technologies must purchase additional hardware and upgrades in order to support an increase in workforce access and data. Mandatory software upgrades are also required to support systems should there be a risk of hardware failure. In addition, businesses that use the traditional approach have their own IT department to install, upgrade and maintain the hardware and software.

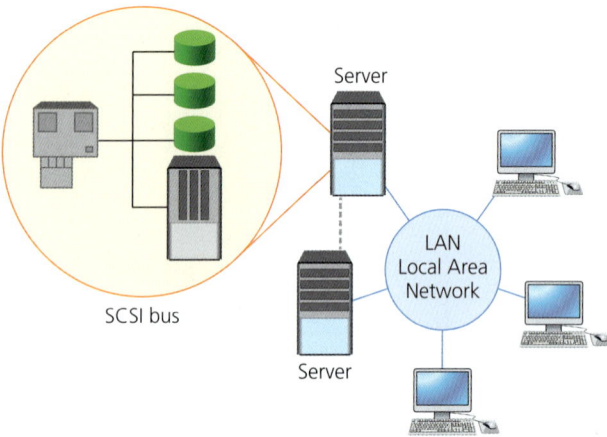

▲ Figure 5.5 A traditional IT solution

The traditional infrastructure as mentioned above is also considered one of the most secure solutions for hosting data, as businesses can maintain full control of their software and data on their 'local' server.

Cloud technology

Instead of accessing data via physical hardware, the servers, software and networks are hosted 'in the cloud' and not on the premises of the organisation. Cloud solutions are a real-time environment hosted between a number of servers simultaneously. Instead of buying servers, an organisation can rent data storage space from cloud computing providers on a pay-per-use basis. This can be more cost-effective.

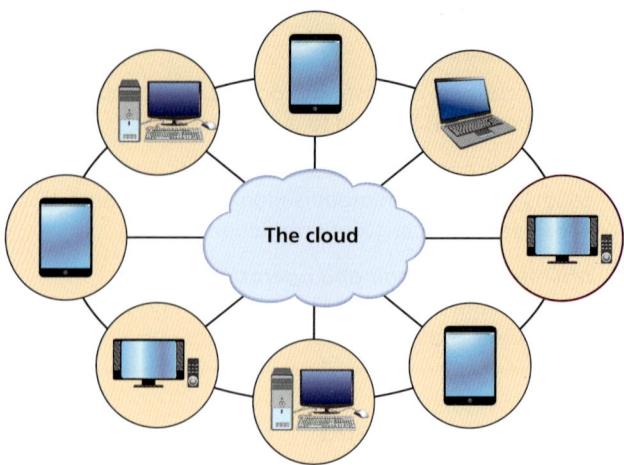

▲ Figure 5.6 The physical hardware, servers, software and networks can all be hosted via cloud computing

The difference between the cloud and traditional IT solutions

Traditional IT systems are not as resilient as the cloud. It is not guaranteed that there will be a consistent high level of performance from the servers. There is limited storage capacity and there can be periods of downtime. This can disrupt the productivity of the organisation.

The information and applications hosted by cloud providers are distributed evenly across their servers. If one of the servers fails, there is no loss of data or downtime as there are still other functioning servers available. The cloud also offers greater storage facilities, server resources and computing power.

With traditional IT systems, an organisation only has the resources that are available to them. If an organisation runs out of storage, then they would have to consider upgrading or leasing another server. Although this can be relatively simple in practical terms, there will be important time and cost considerations for the business to consider.

For example, a business might need to increase their storage capacity for a limited period – meaning that the return on their investment will be smaller – or they might need to increase their capacity very quickly – meaning that they will have to look for quicker, and likely more expensive, options.

Scalability can link to the increase in resources as well as the decrease in resources based on the needs of a business. Consider a business that is expanding quickly and needs to expand its storage. Depending on what systems for storage they have in place, it may be just a case of adding additional drives to the servers. On the other hand, it may require the need to deploy additional servers. Adding additional drives to a server is not as costly or time consuming as it is to deploy additional or even new servers.

If a business is reducing in size, then it may need to 'scale down' its IT resources. Assessing what is no longer required and possible adaptations to processes takes time as well as planning what to do with the redundant hardware/software. This can have cost and time implications as any data needs to be securely removed/destroyed and equipment then needs to be securely disposed of.

Cloud hosting provides more flexibility and scalability. The on-demand virtual space means that there is unlimited storage space and an increase in server resources is also available. Cloud servers can scale up and scale down depending on the needs of the organisation. The organisation can also install software on an 'as needs' basis. Cloud hosting can provide businesses with greater scalability and flexibility, but it is very much dependent on the business' individual requirements. If the provision of PaaS, Saas, etc. are not in line with the business' current systems with respect to hardware, platforms, and software, then processes may need to be adapted by the business first (if at all possible). This can be time consuming and costly.

A cloud hosting facility is managed by the cloud storage provider. They manage the hardware and implement, control and monitor the security measures to ensure the system is running smoothly and efficiently. Organisations that use the traditional IT solution must have in-house management to carry out these functions. This can be costly and time-consuming. The in-house team must include IT professionals who monitor and maintain the servers, including installations, upgrades, configuration issues and threat protection.

In many contexts, cloud computing can be more cost-effective. Organisations only pay for what they use. In addition, decreased risks of downtime means that productivity is improved and therefore the organisation will benefit from an increase in profits. With traditional IT systems, hardware such as servers decrease in value as they age, resulting in little return on investment for the organisation.

Many people think that cloud computing is less secure as data storage and software delivery is external to the organisation. It is important for organisations to research the different cloud providers and select one with a proven track record with the security measures they provide. The vast majority of cloud providers carry out regular maintenance and security of the servers. This includes the maintenance and upgrades of software and even hardware depending on what cloud environment a business is using.

> For more about cloud environments, see Content area 7, section 7.4.

This can save a business a considerable amount of time and money. In the context of traditional IT systems, security of data storage is the responsibility of the individual organisation. Time and money is needed to employ staff with the correct level of expertise and knowledge in order to implement appropriate and robust security data storage measures.

Organisations will still, of course, need their own security systems and staff (an e-commerce site, for example, will need to ensure that its own site is secure in terms of the management of access keys, user privileges, and so on), but working with cloud providers can reduce the overall burden.

Examples of how digital technology can be used to improve business collaboration

Digital technology can assist the collaboration of businesses who want to work together on different projects. The oil and gas industry have used digital technology to work with business partners for the development of oil fields, oil and gas exploration and oil and gas processing. This would include the use of project management software so that all businesses and the project teams within each business can monitor the progress of projects, budgets, timescales and deliverables for each project phase, as well as the tasks that are to be completed and the resources required.

The implementation of the cloud provides accessibility to files and information in real time. This means that documents can be worked on and updated without 'the latest version' being sent to all interested parties who have right of access. In addition, very large documents, audio and/or video files cannot normally be sent via email as there are restrictions in place of the size of files that can be attached to an email. These large documents and files can be stored on the cloud and access provided to those who need them so they are available when required with little or no delay.

Also, the workforce can be asked for feedback and/ or ideas using online surveys created by the company, which gives them a feeling of being valued. This sense of inclusion is good for company culture and can motivate the workforce to work more collaboratively. There are, therefore, many benefits to implementing digital technology to improve business collaboration, whether it is between departments, individuals within departments, other businesses, suppliers and even with the customers.

5.1.4 Factors that determine the feasibility of a digital project

Feasibility study

Part of the project management life cycle is to conduct a feasibility study during the initiation phase. The feasibility study is carried out to determine the viability of a project. It analyses whether the project is feasible and worth investing money and time in.

This means assessing whether it is logistically and practically possible to deliver the project, in the context of a business' resources and other commitments, without contravening legal regulations and requirements. To be economically justifiable a project should deliver a return – be that financial, in the form of profits or savings, or another business benefit, for example a change in how a business operates in line with strategic goals.

A feasibility study can also help to identify any potential issues and/or problems. This highlights risks to the project and its success, enabling the project team

to mitigate them. During the feasibility study, project managers will consider the various resources that are required, including people, budget and technology. They can then ensure that there are sufficient resources available or, if not, discuss further resources requirements with interested stakeholders prior to the project starting.

Feasibility studies also enable a business to look at a project in a wider context. How does the project schedule overlap with other projects and commitments? Will the project put strain on other areas of a business? How does it fit in with a business' digital road map? If a project is likely to impact on other areas of a business, then the relevant stakeholders should be consulted and mitigations should be put in place to limit those impacts.

Some of the benefits of a feasibility study

Benefits of using a feasibility study include:

▶ It allows a cross-section of the business to consider and agree on the project.
▶ It enables the key stakeholders in the project to establish the viability and potential success of the project.
▶ It pre-empts issues and problems, allowing a project team to mitigate or minimise them.
▶ It can help establish whether a project should be delayed until further resources become available, for example, funding to increase the available budget, time allocation, human resources availability including access to people with specialist skills.
▶ It provides a risk/return on investment analysis; in others words, does the return on investment to the business outweigh the risks involved?

Some of the drawbacks of a feasibility study

Drawbacks of a feasibility study might be:

▶ They can be time consuming and therefore there may be insufficient time or expertise to carry out a detailed study.
▶ It is commonly only a documented (written) analysis of the project and while it may include the identification of potential issues and risks, it may not identify problems that can occur in real-time when the project is underway. Therefore, the best feasibility studies will include simulations (where appropriate), to reflect the reality once the project is underway. This is especially the case in projects that are new and unfamiliar to a business.
▶ They can be expensive depending on the industry concerned.

Risks, constraints and dependencies

Risks

Digital projects require the convergence of data, systems and people. Examples of risks include:

▶ Failing to plan – it is very important that projects are carefully planned and minimise any disruption to the functionality of the business while they are being implemented.

▶ Changes to hardware and software availability due to new technology or upgrades can impact a digital project if the changes happen during implementation. This can prevent the new system from functioning as intended or being outdated before it has even been completed.

▶ The organisation merges with another organisation part way through the project and the merger has an impact on the project requirements.

▶ A key project team member or team members leave the organisation part way through the project.

Constraints

Constraints are the limiting factors that can impact on the overall success and quality of the project. Examples of constraints are:

Time

It is essential that the timescales for any project are estimated as accurately as possible. This requires a combination of research and expertise. Time is an important measure for project success.

Cost

All projects have a set budget which must be adhered to. It is important that research is carried out so that accurate costings can be established. Costs relate to:

▶ goods and services
▶ hourly rates of employees working on the project (so accurate timescales are important and can have an impact)
▶ equipment.

Scope

The project scope is a guaranteed set of deliverables. This is what the project will achieve if successful, if the budget and timescales are appropriate and agreed.

Quality

This is one of the major constraints of every project. The quality of the outcome of the project is dependent on the scope, time and costs.

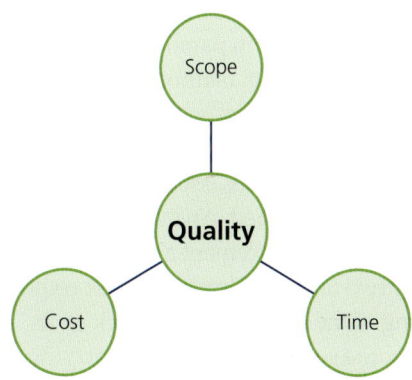

▲ Figure 5.7 The quality of a project's outcome depends on scope, time and cost

Any changes to one or more of the other constraints, that is cost and/or time and/or scope, can impact on the quality of the outcome of the project. In addition, any changes to the expectations of the quality of the outcome of the project can have an impact on time, scope and cost.

Dependencies

It is important to the success of any project that the project's dependencies are set out. The role of the project manager is to:

▶ set the sequence of tasks within the project plan
▶ calculate the time each task will take (this is usually referred to as the task's critical path)
▶ identify the resources required to complete each task as well as identifying any potential scheduling problems
▶ monitor and manage the tasks
▶ identify and action any opportunities to accelerate the schedule of the project's task.

There are four different types of project methodology:

1 **Finish-to-Start (FtS)** – this is often referred to as a finish-to-start dependency. The first task (predecessor) must be completed before the next task can start (successor). Once the task that was classed as the successor has started, it becomes the predecessor to the task that comes after it and the next task in the list becomes the successor, and so on. An example could be a project to install new robotic machinery in a manufacturing company and to train staff how to use it. Leaving to one side the many tasks to be carried out to install the robotic equipment, the training of the staff (the successor task) cannot start until the installation of the robotic equipment (predecessor task) is completed.

2 **Start-to-Start (StS)** – there are some projects that require tasks to be carried out in parallel. These projects will tend to use the StS methodology. Each task still has the predecessor/successor dependency scenario but the successor, while it can be carried out at the same time as its predecessor, cannot start until the predecessor has started. An example could be that within a project to implement a new digital system, there are two tasks involved. One task could be the monitoring of the new system to check on its performance, etc., the second task could be carrying out stress testing to ensure that the system can handle the very large volumes of data it will need to process. The stress testing can be carried out alongside the monitoring of the new system, but the monitoring element has to start before the testing task can begin.

3 **Finish-to-Finish (FtF)** – within this methodology, tasks can still run in parallel, but a successor task cannot complete until its predecessor task has completed. For example, a new network system is being installed along with new network cabling. While the devices can be prepared and tested individually, they cannot be tested as part of the network until the installation of the new network cabling has been completed and tested. Then, the devices can be installed onto the network and network testing can be carried out.

4 **Start-to-Finish (StF)** – this is where the predecessor task can only finish once the successor task has begun. This is not such a common methodology to use and is very much dependent on the type of project. An example would be the installation of a new software programme to be used in a specific aspect of the business. It is to replace an out-of-date system. While the new software is being installed, configured and tested, the old system still has to function to enable the business to continue with its day-to-day activities. The old system cannot be deactivated/removed, until the new system is up, working and ready to function.

Test yourself

1 Explain the purpose of a feasibility study.
2 Identify one benefit and one drawback of conducting a feasibility study.
3 Explain the difference between the FtS and FtF project dependencies.
4 Discuss how constraints can have an impact on the overall quality and success of a project.
5 Describe the three project dependency categories.

5.1.5 How digital is used to meet user needs and ensure quality of product/service

It is important to consider the user needs (internal and external to the organisation) when implementing digital technology. It is part of the overall project planning. In addition, it is also important to consider how the digital technology will ensure and even enhance the quality of the service and/or product(s) offered by the organisation.

Appropriate and effective functionality

Digital technology can improve the way the employees carry out their tasks. This can be in relation to their interaction with colleagues within the same or another department, or with external people such as suppliers and customers. Digital technology can speed up existing processes and/or provide new and more flexible ways of working. In the previous sections of this content area, you learned about the various functional areas within organisations and how digital technology can be used to improve processes. This can also enhance the quality of the services and/or products offered.

Before implementing additional and/or new digital technology, it is important that an assessment is made of the current systems already in place and what is needed to meet the requirements of the organisation and its external stakeholders. It is therefore important to think about the inefficiencies of the current system. The digital technology solution for each organisation will depend on the industry they belong to. It is important for an organisation to research what its competitors are using.

Reduction of pain points

There are several issues that can arise which cause pain to the organisation, its productivity and its employees. Some examples of potential **pain points**:

▶ **Low budgets** – while organisations are under pressure to embrace digital technology, in some instances the budget available for implementing the technology is severely underfunded. The costs can increase quickly and organisations often try to cut costs in an attempt to save money.

▶ **Communication** – a lack of communication or miscommunication between the IT department and other departments in the organisation can be a barrier to adapting to, and entering, the digital technological era. All departments have the same goal – the success and growth of the business. But invariably these departments have different strategies to grow the business and increase its success. It is therefore important that all departments communicate effectively with each other to share their visions and collaboratively improve the business.

► **Lack of technical knowledge by end users** – IT professionals often spend precious time explaining basic IT functions to people because they lack the knowledge to do things for themselves. It can be very time-consuming during meetings to explain in detail how potential new and/or upgraded digital technology can improve the functions of the business. A wider knowledge of digital technology within an organisation can save time and resources.

► **Lack of training** – a lack of knowledge is an issue when implementing digital technology. End users who do not receive appropriate and adequate training will not be able to use the technology effectively, which in turn will have an impact on their ability to carry out the required functions. Good training for the end users will ensure the overall successful implementation of the digital technology, saving time and resources.

► **Security** – this is one of the biggest priorities of any IT department. **Data virtualisation** (although costly) has made organisations more vulnerable to security breaches. Data breaches can have an impact on an organisation through data loss as well as loss of customer confidence and trust. Budget cuts can lead to a reduction in the funding of security of systems.

Key terms

Pain point: issues that people will work around. In some instances the user is not even aware they are happening.

End user: a consumer of a product and/or service. This does not only apply to customers/clients but also to employees.

Data virtualisation: connects all types of data sources regardless of the file types and locations. The data is then combined, and users can access the combined data through reports, mobile apps, websites, dashboards and portals.

Accessibility considerations

The implementation of digital technology must consider the accessibility to the system by relevant users (internal and external to the organisation). The accessibility principles are covered in Content area 4.

Refer to Content area 4, section 5.1.5, for further information on socioeconomic status.

Compatibility

The implementation of new or additional digital technology requires organisations to consider whether it will be compatible with the equipment and systems they already have in place. If the intended additional or new technology is not compatible, then this can result in higher costs to change entire systems and equipment. Restriction in budgets can mean that the implementation must be reduced, delayed or even not carried out at all.

Availability

There are many things to be considered when implementing digital technology, whether it is a completely new system, an upgrade or an addition to an existing system. Availability refers to the expertise of the people who will be implementing the technology and delivering training on the use of the technology, as well as people who will be using it. It also refers to the availability of the technology itself and how compatible it is with the systems and processes already in place.

Good user experience

It is important that when implementing digital technology, organisations and their designers analyse the current user experience. They need to consider the demand, the need and the usage to make future improvements.

Digital technology can be used to enhance user experience in a several ways. Chatbots, for example, provides the customer with the opportunity to resolve their queries at any time of the day or night. On the other hand, they can be infuriating if, after answering the questions posed, the chatbot states that it cannot help and the query will need to be sent to an actual human.

However, for simple queries, they are useful. There is the opportunity to have live chats with a customer service representative for many systems and this is positive for the user experience. They are 'talking' to a human who is usually able to resolve their problem, answer questions, etc. However, some live chats do not allow the user to pay for services or products.

The use of mobile technology has provided customers with an additional means to obtain information about products and services as well as accessing FAQs (frequently asked questions), without having to be at home in front of a computer.

A customer road is useful when planning the implementation of new or extended digital technology. It is a diagram or set of diagrams that show the key stages that customers (internal and external to the organisation) go through when interacting with the organisation. This includes the purchasing of products and services, accessing customer service through the phone, email or social media for external customers, as well as the processing of data and communicating with other departments and so on for the internal customers.

A customer journey map views the business from the perspective of a customer and should draw on detailed and comprehensive customer research. It will enable the business to understand how and when a customer comes into contact with it, across platforms and media and throughout the sales and user journeys, and to improve the user experience accordingly.

It is important to remember that not all customers are the same. A good road map allows for that variation and breaks down the customers into customer groups, or segments. This is particularly important in the context of digital technology, as certain customer groups might be more or less familiar with digital technologies (for example preferring to speak to a customer service team on the phone rather than via a chatbot).

Cultural awareness and diversity

> Diversity and inclusion are discussed in Content area 3.

The reception and use of technology in the workplace can be influenced by several factors, including **socioeconomic status**.

It is important to bear this in mind when implementing digital technology. This is because the availability of technology varies according to an individual's socioeconomic status, which is commonly referred to as the 'digital divide'. Whether someone is working for the organisation or is one of their customers, they can all be categorised between those that have technology (and to what extent) and those who have not. This can result in some people being less familiar with certain technologies and therefore reluctant or unable to use them. New technologies might therefore need to be sensitively implemented or accompanied by training.

> See Content area 4, section 4.2.3 for more information on the digital divide.

> Refer to Content area 3, section 3.1 for further information on diversity and inclusion.

Key term

Socioeconomic status: a measure of a person's combined economic and social status (tends to be positively associated with health).

Test yourself

1 Explain the term 'pain point'.
2 Identify two issues that create the digital divide.
3 Explain the term 'data virtualisation'.
4 Discuss how a customer road map can aid an organisation when implementing new or additional digital technology.
5 Explain why a lack of training of internal end users can be problematic for an organisation when implementing new or additional digital technology.

5.1.6 How the characteristics of end users affect the use and characteristics of digital technologies to access a service or a product

Age

The age of the end user can obviously vary. Younger end users who have grown up with technology and the many changes that have taken place are more likely to adapt to new digital technology than those who are older. It is therefore important that any organisation designing, planning and implementing digital technology considers the wide age range of their end users, whether they are their customers or their employees. They need to ensure that the systems and associated software are simple to use.

Skills

There are a number of demographic factors that can have an impact on the skill level of the end user and therefore their experience (or lack of experience) of digital technology. These include age, location, socioeconomic status and so on. All end users must be enabled to use the digital technology by being assisted in navigating and using it with minimal issues.

Education level

The educational level of end users must be considered by businesses. This is regardless of whether they are internal to the business, that is employees, or external, for example customers. If the education level of the end user is low, then the technology should be simplified for use and incorporate language and interaction that is easier to understand.

It can be a difficult balance for businesses to apply in order to meet the individual educational characteristics for a wide range of end users. If a person has a low reading level it is going to be difficult for them to read any significant amount of information on a website or mobile application about a particular product or service that is available. This reduces the accessibility for this customer. Therefore, it is important that businesses consider using graphics and video presentations so that there is potential for the person to access the same information as those who are able to read more fluently. Some potential end users may have little experience of using or training in using digital technology and so it is important that businesses make using their product, for example an app or website, as simple and intuitive as possible for this audience.

Internal/external audience

An organisation needs to consider all end users. Are they internal to the organisation, for example employees who will use the technology to support the function of the business? An example could be an employee working in a warehouse who will use technology to record the incoming and outgoing stock. The external end users are the customers who are purchasing the products or services. Consider a customer who shops online: they will use the organisation's e-commerce website to select and order the products and/or services they require.

Level of technical knowledge

This can be linked to age, skills and education level. The amount of exposure a person has had to technology and how much they have used it can have an impact on their technical knowledge. For example, a person could be used to using a smartphone to make and receive calls, use apps and social media, but if something goes wrong, they may not have the technical knowledge to solve the problem. In addition, there can be limitations of a person's technical knowledge with respect to the specific product and or service. For example, if a person accesses a website to try and resolve a problem with a specific product, for example a television not displaying the output from a digital box for TV services, they may not be technically savvy as to how to set it up or the associated technical jargon. If it is difficult to trawl through a website to find an answer to the problem or to understand instructions on how to resolve the problem because it has been written with a lot of technical jargon, this restricts access for the user. This is why a lot of businesses include diagrams, graphics and even video clips to demonstrate how the problem can be solved or, as in this example, how the devices should be connected.

Additional needs

This includes people with disabilities such as visual, speech and/or hearing impairments. Organisations need to ensure that any digital technology implemented is accessible for all end users, regardless of their situation.

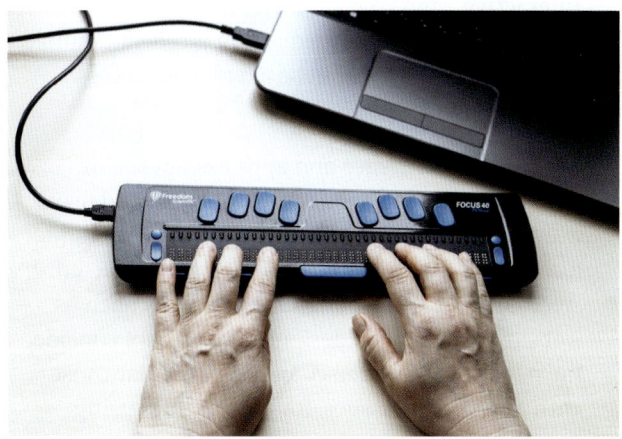

To implement digital technology, an organisation should carefully research the different types of end users, and their characteristics, who access their services and/or products. To do this successfully, **personas** should be created based on the results of the research. Creating personas helps an organisation to recognise and understand that different end users have different needs and expectations. Personas help an organisation to achieve its goal of creating a good and positive user experience for the end users.

> ### Key term
>
> **Personas:** fictional characters, based on research, created to represent the different types of end user that may access the products/services in a similar way.

The positive and negative impacts of digital technology on end users

Positive characteristics

▶ **Across devices** – it is now possible to access content across multiple devices. This has been assisted with the availability of cloud services.

▶ **Mobile** – mobile technology enables a person to be online anywhere at any time.

▶ **Dynamic** – technology is constantly evolving.

▶ **Personalised** – everyone has their own set of 'tools' that they use regularly and therefore create their own digital environment.

▶ **Connected/global** – people can connect on a global scale using social media, email, text messages and online meeting platforms.

▶ **Ubiquitous** (found everywhere) – almost all locations have Wi-Fi allowing people to be constantly online.

▶ **Interactive** – the Web allows for greater interaction than in the past. Social media is an excellent example of how the interaction can be a two-way process.

Negative characteristics

▶ **Insecure** – some sites and apps are not secure (and may even sell a person's data). The end user wants reassurance that any interaction with digital technology is secure and that they are protected.

▶ **Accessibility** – some digital technology such as websites are not well designed and do not consider the needs of the end user and in particular those with accessibility issues, for example not using images/videos instead of text.

▶ **Quantity** – too much information can make it difficult for the end user to find the information they want.

▶ **Time-consuming** – participating in online platforms such as social media has many benefits, but it can also be time-consuming.

▶ **Trivial** – it is important that the information available is useful. It can be too much of a challenge if a person has to filter out trivial information to find the information they want.

▶ **Cost** – resources and tools can be costly.

▶ **Unreliable/Insecure** – sometimes apps, websites and digital technology can crash or get hacked. This has an impact on an organisation as well as its end users. The impact can be operational, financial and/ or reputational.

▶ **Transitory** – technology is constantly changing. Organisations must consider that all end users will have got used to using the current systems and interfaces when suddenly it changes.

▶ **Connectivity** – while it is true that there is almost ubiquitous access, when a connection fails, and end users cannot connect to the systems, it can be frustrating and create problems.

Test yourself

1 Identify two positive characteristics of digital technology.
2 Identify one end user characteristic and explain how it can affect the use and characteristics of digital technologies.
3 Explain why connectivity can be a negative characteristic of digital technology.
4 Describe the term 'end user'.

5.2 Digital value to business

5.2.1 The importance of digital within organisations, and the ways in which digital is used to add value to a company

In many organisations, digital technology is at the heart of their operations. Online retailers and web-based service providers all benefit from the technology available and from ensuring that they stay up to date with technological developments. But there are also organisations that are still developing their digital technology strategy which can require a total change to their existing processes and operations. The benefits of implementing digital technology may not be immediately obvious and can take time.

The world is increasingly connected by and dependent upon technology. It is therefore very important for organisations to adapt to the use of technology and remain relevant and modern. It allows organisations to enhance their performance across the key operational areas, creating added value to their customers. This can be in **business-to-business (B2B)** and **business-to-consumer (B2C)**.

Engagement of customers, users and other stakeholders

Digital technology has provided consumers with a vast array of products and services worldwide. This includes tools and techniques to enable them to research products and services before they buy them. Online reviews such as consumer forums and rating systems on retail sites such as Amazon have ensured that the voice of the consumer has been heard. This also enables businesses to benefit by looking closely at the needs and concerns of their customers.

Consumers can compare products and services from multiple organisations who are all competing for the same business. It is now easier for consumers to voice their own positive or negative experiences with a particular product, service or organisation to a global audience using the internet.

Digital technology has changed the way we all communicate with each other and with businesses. Most people expect to contact organisations using digital channels of communication as opposed to a letter or a phone call.

Digital technology can help the staff of an organisation handle the increase in enquiries and demand more effectively. It is important that an organisation adopts a multi-channel communication strategy so that all customers, users and stakeholders can select their preferred method of communication. This encourages more involvement from all interested parties and reduces the pressure on staff.

By implementing a multi-channel communication strategy, organisations provide their staff with the opportunity to focus their time dealing with large volumes of customer service requests instead of spending a lot of time on the telephone. It also means that there is less likelihood of organisations having to recruit more staff to deal with enquiries. The implementation of digital technology can therefore enhance efficiency and reduce costs.

Providing the opportunity for internal and external stakeholders to interact with an organisation can also improve their experience (remember all stakeholders are customers in one way or another, internal and external to the organisation). Instead of having to make lengthy phone calls which can mean waiting on hold for long periods of time, they can arrange a face-to-face meeting or submit an online request form. These requests can be submitted from anywhere at any time.

In previous sections of this content area, you have learned about the ways that digital technology can be used within organisations. These all add value to an organisation.

Provision of products and services to customers

The arrival of the internet and mobile technology has now made it easier for customers to gain access to a wider range of products and services. Customers are therefore provided with a greater freedom of choice. The increased number of e-commerce websites means that customers no longer have to stick with products, services or suppliers that they are not happy with. They can find and try out new suppliers, and different products and services. The biggest factor for the price and costs of products and services is supply and demand. There is now a greater supply of products due to technology, and businesses are forced to reduce their prices or lose customers to businesses offering a cheaper option. Technology provides added value to the customer by enabling them to spend less money on products and services that would have previously been more expensive.

At one time, a customer would have been forced to wait in long queues to interact with a salesperson when buying a product or service. The implementation now allows customers to buy their desired product and/or service from the comfort of their own home. Technology and the businesses that implement it give customers the control to buy what they want, when they want and how they want. Mobile technology removes the restrictions of location and time.

Social media, blogs and digital magazines are libraries full of information that can be shared with billions of people worldwide. This has led customers to become more educated about what is available. They can access information quickly on a product and/or service by researching a supplier, a product or a service.

Automation technology has rapidly increased over the years and has radically reduced the production costs for businesses. This stems from the production of goods through robotics, AI software and other automated processes. Automation has also extended to customer relationships with many businesses automating their interaction with customers. Technologies such as chatbots, AI and behaviour-triggered tasks are now the norm ensuring that the interaction with customers runs smoothly with minimal human intervention.

Measurable value

Overhead costs

Technology can reduce overhead costs:

▶ **Reduction of mistakes and errors** – the cost of manufacturing is not just about the cost of producing the finished product. Faulty products are a liability to a manufacturing business and as well as having to replace or repair faulty products, there is also the possibility of having to refund the customers (who may not be pleased to have a faulty product). Digital technology can reduce the number of mistakes and errors by implementing technology and automating processes, providing a better customer experience (less likelihood of them receiving a faulty product) and enhancing the overall reputation of the business.

▶ **Reduction in energy costs** – the use of greener and more efficient technology invariably results in a reduction of costs for energy. Many organisations use sensor controlled light switches. The light will not come on and/or remain on unless it can sense someone moving in the location of the light.

▶ **B2B integration** – this is the integration, automation and optimisation of vital business processes that are external to the business. Receiving orders from customers electronically makes it easier to process orders more quickly and accurately. Access to suppliers electronically also makes it easier to view global shipments, automate the warehouse and distribution sites, as well as optimise stock control. B2B integration systems can also reduce or even eliminate the need for human handling for activities such as sorting and circulating emails, preparing documentation and data entry. It also reduces the cost of using stationery such as paper and envelopes, telephones and postage.

Improving efficiency

Businesses use technology within different departments to help improve the efficiency of the business. What technology is used and how it is used depends very much on the type and size of business and the sector in which it lies.

There are main areas within the functionality of a business where digital technology can improve efficiency. These can include:

▶ improved communication between:
  • departments/members of the workforce
  • businesses and suppliers
  • businesses and customers

▶ enhanced collaboration between departments, for example:
  • as part of the production flow
  • when working on projects/innovation and design by sharing knowledge or ideas
  • improved connectivity of the workforce, for example remote workers as well as those situated in other offices or locations

▶ improved monitoring of processes which can include:
  • advanced identification of potential problems/risks, providing earlier opportunities to find and implement solutions.

> ### Research
>
> Carry out research on a business within the travel and tourism industry and analyse how the use of digital technology can:
> ▶ improve efficiency
> ▶ reduce overheads.
>
> Prepare a report of the results of your research.

Facilitating growth

Digital technology helps facilitate the growth of a business by improving the efficiency of business processes, for example by automating tasks, reducing the potential for errors and reducing overheads. In addition, businesses can use the technology to target and reach a much wider customer base, enhance relationships with customers and encourage brand loyalty.

Digital technology also provides businesses with the opportunity to scale their workforce depending on changing requirements. If, for example, a business wants to work on the design of a new product and needs specific expertise (just for the design project), they can recruit additional workers with the appropriate expertise from anywhere in the world as a contractor, freelancer, etc., to work remotely. Once the project is complete, the workforce can be reduced.

Recruiting talent

Digital technology provides businesses with a much wider platform to advertise posts which opens up the pool of potential applicants. An example is where HR can use platforms such as LinkedIn which allows them to target specific people whose profiles suggest they would be of interest to interview for a particular vacancy that is available. Businesses can advertise their vacancies and reach a much wider pool of potential applicants. Depending on the type of business, it is possible that the workforce can work remotely and therefore it provides them with the opportunity to recruit talent from anywhere in the country or even the world. This provides businesses with access to a wider pool of expertise for different areas of their business.

Supporting processes and business models

Sections 5.1.2 and 5.1.3 explained how digital technology can support processes for product research and design, manufacturing control, and remote and local working. The effective use of digital technology for data modelling allows a business to define its core business rules, which results in fewer revisions during implementation of the system. By integrating the requirements and developments of the business model, there is a reduction in the overall development time which means that projects, new services and products can be made available to the consumers more quickly. Data modelling also identifies errors more quickly. It ensures that a business can control the data, reduce costs and accelerate development.

Context and market environment (stakeholders, user profiling, personalised/ appropriate content, data)

Social media can be an extremely useful platform for marketing products and services as organisations can directly engage with their customers and potential customers. Social media also allows an organisation to link with consumers by providing entertaining and informational content that is relevant to its business sector/industry, which does not have to be promotional. Websites that also allow video-sharing provide organisations with the opportunity to share valuable content such as tutorials, information about the organisation and interviews. LinkedIn is a good example of how B2B organisations share their ideas with prospective clients.

Test yourself

1 Discuss how digital technology has improved the engagement of customers with businesses.
2 Explain why B2B integration can reduce overhead costs for a business.
3 Identify one way that social media can be used to market products and services.
4 A business offers life insurance and is considering how they can use digital technology to attract customers. Explain two ways that the business can use digital technology and justify your answers.
5 Describe how data modelling using digital technology enables a business to define its core business rules.

5.3 Technical change management

5.3.1 The factors that trigger change in organisations

Organisational change means a change, re-organisation or replacement of processes, methods, systems, operations, technologies and/or structure of the organisation. This change can be **developmental**, **transitional** and **transformational**. So how does change occur and what causes the change? There may be one or more factors which can create the need for change. These can be financial, economic, technological, social, political, legal, staff related and so on.

Key terms

Developmental: the development of someone, something or even both.

Transitional: the transition (movement) from one position, stage, state or concept to another.

Transformational: producing a change or improvement in a situation.

The factors can be categorised into two groups: planned for factors and unforeseen/previously unpreventable factors.

Planned for factors

Planned change is the preparation of the organisation, either as a whole or a part of it, for new goals or a new direction for the business to go in.

Adding additional features and/or services

This is where an organisation identifies the need to include additional features and/or services such as additional ways that a customer can communicate with the organisation. For example, this might be by using social media or implementing chatbots on the website. It usually relates to improving business processes and the stakeholder experience.

Diversification

This is when an organisation develops a new product or expands into a new market. Usually, diversification is a way that an organisation manages risk by minimising any potential harm to the business during potential economic downturns. Consider how many high street stores now provide online access to their products and/or services due to the lack of customers using the high street to shop. Business diversification is also a strategy for growth.

Scaling

This is implemented to support the growth of a business. It enables a business to grow without, or with at least minimal, barriers. Scaling requires careful planning, finance, appropriate systems, staff, processes, technology and if appropriate partners.

Rebranding

This is the process of changing the corporate image of the organisation. It is usually a marketing strategy of giving a new name, logo or change in the style of a brand which is already established. The purpose of rebranding is to create a different identity within the marketplace.

Adoption of new technologies

As we all know, technology evolves at a fast pace. It is important that organisations adopt new technologies because:

- ▶ **Competitive advantage** – organisations that gain a competitive advantage do not just do one thing. Organisations aim to be ahead of their competitors and to stay ahead by adapting to new technologies and using them in an innovative way.
- ▶ **Avoid possible extinction** – organisations that do not adapt to change, especially in relation to technology, can become extinct. An example of this is when the iPhone hit the market and many mobile phone companies lost their position in the smart phone industry.
- ▶ **Prevent potential financial loss** – the executives within an organisation that do not encourage the managers and employees to embrace new technology can cause the business to lose its standing within the marketplace. This can lead to a loss of reputation, a loss of finance (profit) or even going out of business. Adapting to new technology of course can have the opposite effect and show that the business is a market leader, thereby increasing profits through more customers.

Changes in legislation

These are imposed changes that are mandatory. Organisations must always be compliant when it comes to legislation to avoid severe penalties. Finance and utility industries are very carefully regulated. Organisations may need to reorganise their business processes and systems to maintain compliance with a change in legislation.

Response to competition

It is important that an organisation understands what its customers want and need, and to react quickly, so that its competitors do not gain competitive advantage. To remain competitive, an organisation must ensure that its main focus is on its customers. Organisations must also understand the strengths and weaknesses of their competitors and how well they react to their customer needs and changes within the industry. Organisations need to ensure that they are one step ahead of their competitors while complying with regulation and **competition law**.

> ### Key term
>
> **Competition law:** the purpose of this law is to promote healthy competition. It makes it illegal for anticompetitive agreements to be in place between two or more organisations, for example to share markets and fix prices. It also makes it illegal for businesses to abuse their dominant market position.

Unforeseen or previously unpreventable factors

Crisis

This is when bad things happen requiring all kinds of changes by an organisation. This can include natural disasters such as floods, terrorism and cyber attacks. Many of these are covered in Content areas 4 and 8.

> Refer to Content areas 4, section 4.1, and Content areas 8, section 8.1, for further information.

Zero-day vulnerabilities

This is a vulnerability in a system or device that has been identified but not yet resolved. Somebody who makes use of and benefits from a zero-day vulnerability is called a 'zero-day exploit'. Zero-day vulnerabilities pose a higher risk to organisations and users because they have been discovered before security researchers and software developers become aware of them and can issue a patch. Cyber criminals are quick to exploit these types of vulnerabilities and vulnerable systems are exposed until an appropriate patch has been issued.

Data corruption

Damaged files and/or corrupt data can be inevitable and not a matter of 'if' it happens but 'when' it happens. There are several possible causes of data corruption:

- ▶ malware/virus infections
- ▶ sudden loss of power forcing a power shutdown

- voltage spikes
- physical hardware issues
- bad program exits
- any interruption in the normal processes being carried out by the IT system
- over-sized databases
- network transmission issues, for example attenuation, where there is a loss of signal strength. Within WiFi technology this can occur as the device is moved further away from the router. With wired networks it can be as a result of signal loss within the network cables and the connectors. Potentially, any degradation in the quality of the signal can result in transmission errors and the corruption of data. Packets of data can be dropped, meaning that although they are transmitted, they do not arrive at their intended destination.

> **Research**
>
> For each of the possible causes of data corruption listed above carry out research and explain how each of the causes could be prevented and/or the risk of data corruption mitigated.

System failures

System failures can occur through hardware and/or software faults and/or cyber-attacks. There is also the possibility for the system to fail due to human error. When a system failure occurs, it may not always display an error message and the system may freeze, reboot or stop functioning altogether. Any system failure can cause major issues for a business and can result in it being unable to function until the problem is rectified. This can take time, while the reason for the failure is investigated and a solution is identified and implemented. The results of the investigation into any system failure can trigger an unforeseen change in a business. It can result in:

- new equipment and/or software having to be sourced and installed
- legal, ethical and moral repercussions of a cyber-attack such as a restriction in operations leading to customer dissatisfaction, for example if there is a system failure with a bank, then its customers may not be able to pay their bills or access their money
- changes to operating processes
- additional staff training
- loss of business, either because of downtime or reputational damage.

The types of change that are triggered within a business and the timescales available to implement the change(s), depends on the severity of the system failure.

> **Test yourself**
>
> 1 Describe the difference between developmental and transitional.
> 2 Identify two causes of system failure and explain why they may be unforeseen or previously unpreventable factors that trigger change in organisations.
> 3 Describe the term 'zero-day vulnerabilities'.
> 4 Discuss why it is important for companies to consider triggering change in response to competition, rebranding and diversification.
> 5 Explain competition law.

5.3.2 The technical change management process

The purpose of technical change is to replace or upgrade technology for a business benefit, for example to improve productivity, facilitate growth or increase profits. Technical change can be complicated, costly and disrupt normal business operations and processes. Any change should therefore be carried out with the goal of minimising that disruption and pre-empting problems or issues before they occur.

The technical change management process, which is part of the ITIL (IT Infrastructure Library) framework, is the standard way to approach and implement technical change across industries. This process is a set approach for reviewing, authorising and managing technical change.

The digital change management process is usually led by a digital lead, known as a digital change manager, although the rationale for the change is initiated by senior managers and relevant stakeholders, for example the board of directors and sponsors.

A change request is submitted to the owner(s), stakeholders or sponsors. They will liaise with the Change Advisory Board (CAB) who will advise whether it is feasible to approve the change. The CAB consists of an expert technical team who consider the technical details relating to the potential change and an expert business team who consider other risks, finance, resources, etc. Once the CAB has provided the relevant major stakeholders with the feasibility study of the potential change, it is either approved or denied. If the requested change is approved, the planning and scheduling for the change takes place to ensure that all finances and resources (hardware, software, time, workforce) are in place and readily available. In addition, they will consider how to ensure that the business can still function as intended while the change is taking place and how to mitigate any potential risks to production, equipment and data. Once the planning and scheduling have been agreed, then the digital change is implemented and documented. Along with the implementation, testing will take place to ensure that the change is working as intended and that it does not impact on other processes, etc. Once this has successfully been achieved, then the project is closed.

There are three main categories of technical change management as defined by ITIL. These are:

► **Standard:** these are known changes to services and/or IT system infrastructure where the risks and impact on the business are low. They are pre-authorised by senior management and can be repeated on a rolling basis without need for re-approval. These can be tasks such as installing a new printer or computer to a network, or virus and software updates.
► **Normal:** these are changes that have to be approved prior to being carried out. Changes that are high risk also have to be approved via the CAB. This can be for things such as installing a new server, installing new network cabling, etc. The business still needs to be able to function while these changes are being implemented. Changing, adding servers or installing new network cabling can result in system failures or systems being out of operation while the current systems are being transferred across. This therefore can have an impact on the business' ability to function for a period of time, which may be hours or days.
► **Emergency:** these are critical changes that must be carried out as a matter of urgency. A typical example would be in response to a cyber-attack on a network. These changes cannot be planned for.

Identifying the changes to be made

Initially, the intended changes need to be identified as well as the reasons behind the changes, the risks involved and anticipated timescales. It is also

important to consider in what order the changes should be implemented.

Identifying and communicating potential risks and the desired impact(s) to stakeholders

When identifying the changes, the risks involved are also considered. These must be communicated to all relevant stakeholders (internal and external to the organisation). The communications must be positive and timely. It is important that the internal and external stakeholders, including end users, are confident in the changes being made and that any changes will not create too much disruption to them. Communication can include online meetings, email, website and social media postings.

Configuration of the new system and process

Careful planning of the changes to be made must be clearly documented so that the people carrying out the changes can implement them in the correct order, using the correct equipment, configuration requirements and adhering to the identified timescales. All changes must be tested to ensure that the systems and processes work as intended.

Method of implementing change

There are several different methods of change. These are:

► **Parallel** – this involves operating the old and the new system simultaneously over a period of time. This ensures that any major issues can be identified and rectified without any loss of data, functionality and/or production. This also allows end users to familiarise themselves with the changed system/process.
► **Phased** – this is where changes are made in stages, that is not all changes are made at the same time. This is to ensure that the changes are functioning as intended before implementing further changes and, as with the parallel method, ensures that there is a reduced risk to data loss, operation and production. The phased method can also be used in conjunction with the parallel method.
► **Direct** – this is where the changes are implemented directly into the system already in place or a new system/technology is installed but is not phased or parallel.
► **Pilot** – the pilot is used to test the change deliverables as they are being used. This is to address any problems which can be fed back

to the technical working group to address. The pilot is usually carried out by end users who have experience in using the current system. Microsoft would often ask people to trial their new version of Windows (beta versions) and feed back any issues they identified.

Documenting the change process

Change management reports are required by all organisations to document the changes. These are used to monitor and/or control the changes. Documentation will differ from organisation to organisation, but commonly includes:

▶ **change request form:** a document used to request a change and outline the impact on the business. This will be considered by the CAB in the first instance.
▶ **implementation plan:** after approval, a plan will be drawn up to define how the change will be implemented and how any risks will be mitigated against.
▶ **decision log:** a record of decisions made during the implementation of the change.
▶ **test plan:** a document detailing the testing that will need to take place, including timelines, before launch.

Importance of rollback planning

This is the most important aspect of any change management process to ensure that there will not be any impact on how the organisation functions. The rollback plan helps to prevent potential downtime as data can be restored quickly from a backup if problems should arise. The plan will include the detailed steps to be followed should there be a need to roll back the changes to the state before the changes were made.

Importance of ensuring reproducibility of performance and outcome

The intention of implementing any change is to obtain improvements to production, performance and services. The improvements can be in relation to the development or increased production of goods, employees using the systems and equipment working more effectively, and a better overall customer service experience. It is therefore important that there is something to measure against (key performance indicators – KPIs). There should be a detailed analysis of the performance and the outcomes the organisation wants to achieve prior to any changes so that the performance and outcomes after the changes can also be analysed. The performance and outcome should at least be the same. The last thing an organisation wants to do is implement change which reduces performance and outcomes.

Traceability of requirements throughout the development life cycle

Requirements are traced throughout the life cycle including during testing and resolutions of issues. The requirements are traced backward to the source of the requirement, for example to comply with legislation, to increase productivity or to improve the customer experience. It confirms that the requirements have been met and can also accelerate development time. It is also important for analysis purposes, for example if there is a change to a requirement, then the traceability can determine what the impact of that change would be. Many organisations will create something called a traceability matrix. This is used to create an audit trail which is a mandatory requirement for regulated sectors, for example healthcare, oil and gas.

Test yourself

1 Explain what a rollback plan is used for.
2 Describe two different types of changes that can take place.
3 Discuss the methods that can be used when implementing change. Include examples of which method would be best for different types of industry.
4 Identify three documents that are used within the change process.
5 Describe how traceability works.

5.3.3 How organisations respond to, prepare for, manage and reinforce change

Responding to digital change

How you respond to digital change depends on the organisation, but there are key steps that should be followed.

▶ **Visualise the digital journey** – every digital journey requires a plan which initially requires a certain degree of imagination. It is important that the organisation imagines how the digital strategy will impact on the business and considers how it will impact on the culture of the organisation. In addition, the organisation needs to consider how any digital change will impact on the customers' perception of the business.

- **Take risks** – it has been proven that organisations that achieve a high degree of success within their respective markets take risks (some of them big risks) by implementing new digital technology and innovations regarding the service they provide to their customers, the working environment of their employees, how products are made and so on.

- **Be agile** – organisations that are successful when implementing digital transformation not only think big, but they also act quickly. They become the trend setters instead of the trend followers. It is important that, to be successful, digital strategies are not implemented at a pace which is too slow to adapt and so rigid that they cannot be adjusted.

- **Create a digital culture** – decisions made within an organisation about digital change can influence the entire organisation. It can change the way offices and employees work. In addition, there is a requirement for people with digital knowledge and skills to implement these plans. Professional IT specialists are in high demand and low supply, so this can result in an organisation's plans being delayed.

- **Be proactive throughout the process** – any digital change should be monitored throughout the entire process. The monitoring should consider how the digital change is meeting the goals and ambitions of the organisation, for example how are customers responding to the digital change and how are non-IT employees adapting to the changes? In addition, if the digital change strategy requires some adaptations, then they should also be carried out. As with any change strategy, any technological disruptions, any miscalculations and/or changes to the marketing/production environment can have a negative impact on an organisation's digital change.

Preparing for digital change

Any digital change can be a very complex process and therefore careful preparation is required to ensure that the digital change is successful.

Choosing the appropriate digital change strategy to meet the goals of the organisation

It is important for organisations to identify where digital change is needed and not invest in digital technology just for the sake of it. It is important that organisations carefully consider the overall business goals and objectives they want to achieve. This includes short term, medium term and long term, followed by an assessment of what digital technology is required to achieve the goals and objectives that have been identified. An example could be that an organisation has a goal to expand into new markets. It is therefore more important to focus on the technology required, for example to implement a solid cloud infrastructure to support processes from multiple locations as opposed to investing in AI technology to trial a new production idea.

Investing in technology

Digital change means different things for different businesses in different contexts. Spending a lot of money on digital technology does not always guarantee success. Research has shown that there are three forms of digital technology that can have an impact on the operational functions of organisations:

- the IoT – because it can provide **operational intelligence**
- the cloud – because of its scalability
- big data analytics – which can transform vast amounts of data into predictive and actionable information.

> ### Key term
>
> **Operational intelligence:** this is data relating to the operational performance of the business that is captured in real time. It enables a business to analyse its operations and make them more effective. The analysis of this real-time data is usually automated and therefore is immediately available for consideration by relevant personnel within the business.

It is therefore important that organisations invest in the digital technology to meet their specific needs.

Convincing the stakeholders

Once an organisation has identified how digital change can support the overall business goals, it has to convince the stakeholders. This is because digital change strategies change the way businesses function. There is an impact on people's jobs, how they work together and how they complete tasks. Engaging the stakeholders is not always an easy task for organisations. All staff from the boardroom to the 'shop floor' need to believe that they have a personal and professional interest in the changes being made. It is important that they can understand the reason for the investment so that there is less resistance to new processes. This is extremely important when digital technology is being used to automate processes that would otherwise be carried out by people or when the investment in technology will deliver a profitable return.

Using data to enhance the decision making

It is well known that organisations collect huge amounts of data. But it is important that the data is used to provide insights into the industry, guide the changes that need to take place and even identify new revenue opportunities. Data analytics provides organisations with critical information relating to customer trends, market predictions and how products/services are performing.

Re-evaluating the digital change strategy

A digital change within any organisation is never complete as new technologies are launched continually. This can be robots to complete production line tasks quicker than humans. Or machines which can solve equipment issues without human intervention. It is therefore important for organisations to constantly adapt the digital change strategy as new possibilities arise. They need to reassess the digital journey being taken and consider the rate at which the digital change is taking place. Does it meet customer expectations, does it meet the business goals? If the answer is 'no', then it is possible the digital technology needs to change.

Managing and reinforcing digital change

When managing and reinforcing digital change, there are a number of key concepts that should be considered by organisations.

Setting up vision meetings

The vision is the future aspirations of the business and what it hopes to achieve. A business, for example, might want to move to a fully automated warehouse, migrate all its IT systems to the cloud or enable 50 per cent of the workforce to work remotely.

An initial vision meeting is held to ensure that all relevant interested parties, for example managers, workforce, stakeholders, etc. understand the business' vision and any impact it may have on the way they work or the working relationships with their clients, customers or suppliers. Regular vision meetings are then held to facilitate the development of a shared vision by all parties and to identify any changes that may need to be made as it progresses through the change cycle.

Designing the digital change road map

It is important that any form of digital change is communicated to all interested parties across the business. This includes the customers and the workforce. The customers need to be made aware of any digital change that is going to have an impact on how they secure products and/or services, communicate/interact with the business, how their information is being securely stored, how payments will be processed, etc. By making customers aware of how any digital changes will impact on how they interact with the business, it is hoped that this will retain customer confidence. As far as the workforce is concerned, they will need to know the impacts these changes will have on their job role. This includes whether their job will change, that is what it means as far as the tasks they do on a daily basis. Will there be a requirement for training in order to upskill, will they still have a job? It is therefore important that the workforce is made aware of any changes and to provide them with a degree of confidence in the direction the business is going and what it means to them personally.

Developing the teams and resource acquisition

This is the stage where the main team members are identified and recruited. Each member of the team is selected based on their expertise and knowledge. The team members can be internal to the organisation or hired specialists, for example software developers, systems analysts, engineers, depending on what the change is for. Resources also include things such as hardware, software or equipment that may be required to support the changes. The teams will attend meetings and provide feedback on how their particular tasks are progressing, any issues identified and how they may need to be addressed (or were addressed). These meetings ensure that any digital change remains on track and still meets the goals of the organisation.

Launching and monitoring

The previous sections relate to the key considerations that must be made when carrying out any digital change transformation. Time needs to be given to each of these considerations prior to launching the actual digital transformational change. Once all considerations have been analysed, considered, adopted, then the digital change process can be launched.

Once the digital change process has been launched, it should be monitored closely to ensure that it is still aligned with the documented considerations. There is always the possibility that further considerations materialise and things need to be adapted or amended to address any potential problems/issues. This careful monitoring will ensure that the organisation vision can be achieved even if the end result has to be slightly adapted.

Activity

Divide into four groups. Each group is to select two contexts (no two groups should select the same contexts):

▶ economic, financial and banking
▶ environmental
▶ legal
▶ people
▶ political
▶ regulatory
▶ social
▶ technological.

Prepare a presentation for each of the selected topics, explaining how digital change can be planned for and managed in that context. Present the results of your activity to the rest of the group.

Test yourself

1 Describe two key concepts that should be carried out when managing and reinforcing change.
2 Discuss how organisations should respond to digital change.
3 Explain why it is important for organisations to adopt an appropriate digital change strategy.
4 Identify the key stakeholders to be considered in a car manufacturing company when planning for digital change. Justify your answers.
5 A high street business is considering setting up an e-commerce strategy to increase its customer base by accessing a wider and more global audience. Explain how the organisation should respond to this digital change.

5.3.4 The benefits and drawbacks of technical change in organisations

There are numerous benefits and drawbacks to implementing technical change in organisations. You are going to look at them in relation to several different aspects. It is important for businesses to use technology in a balanced way so that the benefits outweigh the drawbacks. Many of the benefits and drawbacks can have an effect on:

▶ productivity
▶ communication
▶ security
▶ replacing existing products
▶ updating or changing processes
▶ support for stakeholders
▶ costs
▶ stakeholder experience
▶ company reputation.

Productivity

Technology can be used to automate repetitive processes. This can be in relation to the production of goods on a production line, controlling warehouse processes, sending emails and entering data. These types of automated processes can allow employees to focus on other aspects of the business, such as developing new customer relationships or providing a better customer experience.

Technology allows businesses to produce higher quantities of products, making them more consistent and cost-effective.

The use of technology to automate aspects of the production process increases productivity – this allows businesses to reduce prices to remain competitive with other businesses or to increase their profits.

▶ **Quality** – it is important for a business to be consistent with the quality of its products. Automating aspects of the production can help this.
▶ **Flexibility** – it is important for organisations to find a balance between technology and human interaction. Automation is obviously good for mass production, but it is not appropriate for products that are 'tailor-made' to meet a customer's personal requirements.

When employees are provided with access to technology to carry out their jobs, they become more efficient and engaged, and therefore they are happier. When a higher quantity and quality of work is produced by employees, it improves the income made by the business and therefore increases profits. In addition, happy employees are also more likely to remain loyal to the organisation and not leave to work for another organisation. This reduces overhead costs. Technology can be very expensive to buy and to implement. That is a drawback, but it can reduce the cost of producing products. Also, technology can be used to automate dangerous tasks. These tasks are usually carried out by people who receive higher wages because their job roles are dangerous. Again, while the initial cost of the technology is expensive, it can reduce the costs of wages and improve employees' health.

Communication (with stakeholders internal and external to the organisation)

Technological change can have both benefits and drawbacks when it comes to communication with internal and external stakeholders. The range of communication channels now available has increased over recent years and will, no doubt, continue to increase as technology advances.

The availability of a wider range of communication channels provides the opportunity for faster communication to take place between colleagues and between organisations and their customers and suppliers. Even when communicating with people elsewhere in the world, messages can be sent and received quickly and, when the recipient is available (due to different time zones), they can be dealt with more promptly. For example, a customer can access a company's website to search for information, download user instruction manuals for equipment they have purchased, download drivers for their devices, order and pay for goods, use chatbots for information or look up FAQs (frequently asked questions), as well as send communication directly to customer support. This can all be accessed at any time of the day or night that is suitable for the customer. To a certain extent, this means better customer support and, therefore, happier customers and enhanced brand loyalty.

Work colleagues in different departments and areas of the business can communicate and collaborate with each other more easily and more quickly. If work colleagues were required to collaborate with each other and were not in the same location, without access to various forms of digital technology to communicate, there would be time delays while people travelled to meetings, etc. Something that saves time can in many instances save money. Similarly, remote working has become more accessible to employees in certain sectors due to technological change. Employees have access to files and

documents directly from the company system and the availability of the communication channels means that they can still have regular contact with their work colleagues, even when working from home.

While technology can improve the communication between the workforce, it can also reduce the interpersonal working relationships within the working environment. This is because most of the communication is conducted through technology as opposed to having face-to-face discussions. This can have a negative impact on team dynamics, as it can prevent the participants from sharing and exchanging ideas. This is because people wait for audio cues and may not get the opportunity to 'jump in' with their ideas. This can also result in unequal contribution from all parties (it depends on who can make themselves heard the most). There can also be network failures or poor network signals which can disrupt the meeting, and, in some instances, it may not even be noticed that a participant in a meeting 'has gone missing'.

Communications with stakeholders via a digital platform can also have a negative impact on customer relations. The personal touch is lost if all that is available to a customer is communication via live chat, chatbot or email.

> **Research**
>
> Research how technical change has influenced communication between organisations in the healthcare sector and its internal and external stakeholders. Prepare a report discussing the benefits and drawbacks of technical change on the communication methods used.

Security

When implementing technical change, it is important that consideration is given to the benefits and drawbacks with respect to the security of an organisation's systems and, of course, its data. Some technical changes may be beneficial in terms of security. For example, a technical change to secure systems and data from potential cyber-attacks by using multi-level protection methods such as VPNs, firewalls, anti-virus or anti-spam software to mitigate against DDoS attacks. It could be technical change to centralise systems for better control over data and processes and to improve the security of business-sensitive information such as business decisions.

However, the technical change process can also have negative impacts on security and make systems more vulnerable. Therefore, at all times there must be risk management considerations with respect to the security of the systems and data. Throughout the change process, these must be protected from any form of disruption, modification and destruction. For example, if there is to be collaboration between two or more businesses, technological changes might have to be implemented in order for the partners to share data and information. In this instance, consideration must be given to how the security of the data and information is maintained by all parties. While an organisation may readily consider the risks involved and secure their own systems, how can they be sure that the partner companies will protect their data, information, technological knowhow, etc. in the same way?

When implementing technical change, it is important to consider:

▶ Does the change have an impact on the security of the organisation's systems, processes, data and/or information?

▶ Is the impact a benefit or a drawback? If it is a drawback, what other changes are required to mitigate any risks?

▶ Have all aspects of the technical change been considered with respect to security? It may be that there is identification of potential security issues during the actual implementation of the change.

▶ Have any technical changes been clearly communicated to all relevant stakeholders? They need to understand what the change is, the purpose of the change and the timeline for its implementation. These stakeholders will need to be vigilant and consider any potential security issues as well as confirming any benefits and communicate any issues to the technical change management team.

Updating or changing processes

While the benefits of updating/changing processes can lead to improved and more efficient production or response times to customers, it can also have the drawback of time being taken to implement and adjust to the new processes, and potential staff training. For example, if a new automated digital system is implemented to process hydrocarbons that have been extracted from an oil well, then the benefit is that the process can become more consistent and run continuously. The drawback would be that the workforce will require training on how to use the automated system and monitor its performance.

Support for stakeholders

In order to make any technical change successful, it is important to secure the support of the stakeholders. Stakeholders can be external to the business, for example customers, suppliers, sponsors, or internal to the business, for example the workforce. It is human nature to be concerned with any change and different stakeholders will have different concerns, for example the workforce will have concerns about the potential impact to their employment or job roles. Customers may be concerned that they may not get the same level of service or interaction with the business. The benefits of achieving stakeholder support can include for example, additional financial backing from sponsors to implement the technical change or positive response to supporting and applying the technical change from the workforce. Drawbacks can be from the lack of support from key stakeholders who do not accept the change or are against any change, for example customers taking their custom elsewhere or members of the workforce taking strike action.

Costs

The implementation of technical change is not necessarily a cheap and quick fix for any business. There is always a cost implication and what these costs are will depend on what the technical change is. For example, if it is technical change in relation to the installation of a fully automated robotic process on a production line, then this is a costly process to implement. However, the benefit is the amount of money that can be saved over time with respect to production costs once it has been implemented. Therefore, it is important that a cost–benefit analysis is carried out to see if the benefit to the business is worth the cost incurred to implement the change.

Stakeholder experience

As previously mentioned, stakeholders can be internal and external to the business. When considering the benefits to external customers, the business needs to think about their user experience based on the technical change, for example is it easier to pay for goods and/or services, can the goods be despatched to the customer more quickly, can the productivity be increased to meet the demands of the customers? The drawback can be whether the target audience will be able to adapt quickly to the changes and what support needs to be put in place to support them during the change. For internal stakeholders, for example the workforce, do the technical changes support their work activities or does it hinder the work because there are additional processes that they need to follow?

Company reputation

Technological change can have benefits and drawbacks to the reputation of a company. For example, if the communication channels are improved with the customers by providing more access points to interact with the business, then this can enhance the company's reputation. If there are changes to production which increases the production of products, then they become more readily available to the customer, again potentially having a positive impact on a company's reputation. The same applies to automating despatch systems such as the selection and despatch of products from a warehouse.

Obviously, security of information is a major concern and should any technological change result in the loss of data as a result of a system failure or cyber-attack, then the consequences include a loss of company reputation, especially where the issue surrounds personal information relating to customers, suppliers and the workforce. Technical change always has a risk of system failure even with the most stringent mitigation techniques deployed. A system failure can , for example result in the company not being able to function as intended and result in customers not being able to access services or purchase goods, or there could be a delay in paying suppliers. Consider again the situation with a bank, if it has a serious system failure and the customers cannot pay their bills or access their money, this will almost certainly be extremely harmful to the reputation of a company.

Case study

A manufacturer of tyres uses digital technology on the production line, as well as to process orders from customers. They have a research department which has been looking into the use of a different tyre compound. A project team has made a recommendation to upgrade the current technology for tyre production and implement new technology to automate the warehouse. They also intend to use social media to promote the new tyres and inform potential customers of the quality of the tyres, which are not as expensive as the tyres currently available.

Use research to identify potential benefits and drawbacks for each of:
- productivity
- security
- replacing existing products
- updating or changing processes
- support for stakeholders
- costs
- stakeholder experience
- company reputation.

Test yourself

1 Explain why smaller organisations are more vulnerable to cyber attacks.
2 Discuss the benefits and drawbacks of digital technology in relation to internal and external stakeholders.
3 Describe the term 'flexibility' in relation to production.
4 Explain the term 'social listening'.
5 Describe one benefit and one drawback for the use of technology with respect to communication.

5.4 Risks in a business context

5.4.1 The potential risks to organisations of use of digital systems and technologies

Throughout this content area we have looked at how digital systems and technologies can support organisations and the challenges that they face. Many of these challenges are based on the risks involved. We are now going to look at some of them in greater depth.

Security breaches

A security breach is unauthorised access to digital systems and technologies, including data. As you have read previously, this can have a severe impact on businesses as data can be lost, changed and/or stolen, systems can stop functioning and businesses held to ransom until they pay the cyber criminal.

There are many different security breaches that can have an impact on a business. Even when businesses show due diligence by securing their systems and ensuring that up-to-date logical and physical protection methods are in place, there is always a risk that a breach can occur. This can happen by accident or by the stealth of a cyber criminal.

For example, a distributed denial-of-service attack (DDoS) is a cyber attack inflicted on a business and can be targeted at its servers, network and/or website. The target is overloaded by internet traffic with the intention of causing the system to crash as it cannot cope with the vast amounts of traffic. This can obviously have a severe impact on a business, for example customers cannot access the website and therefore there can be complaints, loss of customers and impact on the overall business.

Research

Research the different ways a security breach could occur when digital systems and technologies are used by businesses and the risks that these create. Prepare a report or a presentation of the findings of your research and an analysis of the risks.

Privacy breaches

A privacy breach can occur when a person or people have access to information without permission. Data can include information which is personally identifiable such as name, address, National Insurance number, credit card details, passport and/or driving licence information.

Research

Research the different ways that privacy breaches can occur due to organisations using digital systems and technologies. Your research should include office-based employees as well as those that are remote workers. In addition, consider the different ways that cyber crime can take place. Produce a guide for organisations and their employees to help them understand how privacy breaches can occur and what should be done to mitigate the risks.

Regulatory and legal non-compliance

In Content area 4 you learned about legislation and regulatory requirements. It is important that organisations know and understand the legal and regulatory requirements that apply to their business and situation. They need to understand the consequences of non-compliance and the impact that non-compliance has on their stakeholders (including customers) and the reputation of the business.

Refer to Content area 4, section 4.1 for further information.

System failure

A system failure can mean any aspect of a digital system such as hardware and/or software not functioning as intended. Depending on the severity of the system failure, it can have severe consequences for a business. For example, a business that is totally reliant on its digital systems and technologies in order to function would have a catastrophic effect on a business. If it cannot function, this can have negative financial and reputational impacts and possibly the loss of data, which can have legal implications.

Depending on what part of a system fails, there could also be risk to human life or injury. An example would be an automated process control system in a nuclear power plant. The creation of nuclear energy has to be carefully controlled as the large quantities of radio-active waste is extremely dangerous and has to be monitored. If there is a system failure, this could result in dangerous quantities of radio-active waste being released into the environment, with severe and possibly catastrophic risk to human life.

Audience exclusion

Not all of the world's population can use digital technology or has access to the internet and technology. There are many reasons for this, which include socioeconomic, disability, age and location. When looking at target audiences and the implementation of digital technology, it is important that organisations consider whether there is a risk of excluding any of their target audience. If there is a risk, they need to consider what impact this would have on the business, both financially and on its reputation. The organisation needs to consider how the audience becomes excluded and whether there is anything that can be done to mitigate the risk of exclusion.

Emerging rival services

Technology is forever evolving, and organisations cannot afford to implement digital systems and technology and forget about it. There is a constant flow of technological devices, systems and services being made available. An organisation can become outdated very quickly if it does not continually embrace and make use of the new technology available.

Mobile devices

It is important for organisations to stay up to date with industry trends. The world we live in is moving very much in the direction of mobile devices. It is therefore important for businesses to research mobile trends. This is especially important if the organisation already has a mobile presence using a mobile site or app. People who use technology also like using new and updated technology. Businesses are at risk of losing their target audience if their use of digital mobile technology becomes outdated. In addition, organisations must consider the risks involved with the implementation of digital technology within the processes used in the business. We have discussed several risks of mobile technology as well as the benefits; an organisation must consider all these factors when implementing any mobile functionality, device and/or system.

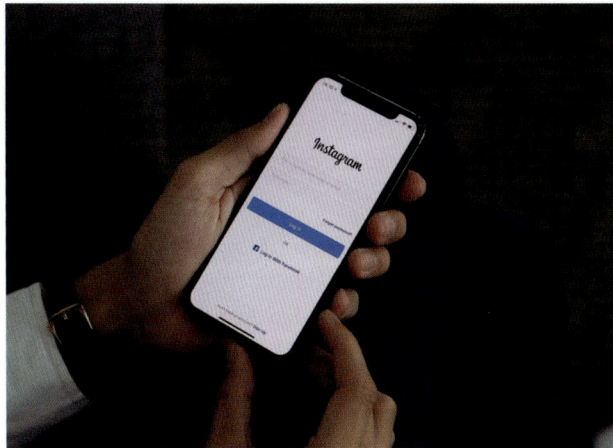

Digital download

Most software is now available through digital downloads via the internet. There are a lot of rival companies offering their own versions of software at varying prices to attract customers.

As with all things downloaded through the internet, there is a risk. Businesses have to ensure that they purchase their software from reputable businesses and that the websites facilitating the downloads are secure. It is not uncommon for digital systems to be infected with viruses or malware if they have been downloaded from suspicious sites.

With digital downloads in relation to software, the business will pay for the software and be forwarded a link via email to access the download. The links themselves can pose a risk as a means for cyber criminals to access a system. The digital download companies save money by not having to deliver physical products in the form of CDs to their customers, but there is usually not a means to confirm that the software has been delivered. Therefore, customers may claim not to have received the software (when in fact they have downloaded it) and claim their money back via the **chargeback** system.

Digital download businesses also compete strongly against each other and try to secure the majority market share by launching their products and services ahead of competitors. The digital music download business is a good example of where the competition is fierce, with businesses such as Amazon Music, Spotify, iTunes and many others going head to head.

Cloud services

In section 5.1.3, you learned about the use of traditional and cloud-based technologies. A cloud environment offers greater speed and flexibility as well as possible cost savings, but there are risks.

A key risk is failing to stay up to date and falling behind competitors. A business needs to be constantly aware of new cloud technologies and the benefits they offer, as missing out on these will give their competitors who are using these new technologies a competitive advantage.

It can be difficult for organisations to evaluate cloud market as cloud providers tend to report their cloud businesses differently. While there are the well-known vendors such as Google, Microsoft, Amazon and IBM, there are other vendors appearing all the time. Some organisations are linked to more than one provider, depending on the cloud service they require, for example IaaS (Infrastructure as a Service), SaaS (Software as a Service) and PaaS (Platform as a Service). Some of the larger vendors have expanded their services to incorporate all three areas (as well as other areas such as hosted private, private, public and hybrid cloud).

For businesses to make informed decisions about which vendor or vendor(s) to use, they have to think of the security of their own systems and data. They will consider the 'track-record' of the different vendors, especially with respect to security, as well as the overall costs involved. Cloud provides will use marketing techniques to entice customers to their products, which can include offering special pricing, for example first year at a vastly reduced rate, as well as promotion of how secure their provision is in comparison to other providers.

Rapid changes in technology and trends

Education

Changes in technology are altering the dynamics of education, especially between teachers and students. The fastest trends are the use of laptops, tablets and other mobile devices. Mobile devices are used as a substitute for handouts, books, paper and pens, and now they are used to change how teaching and learning takes place.

In spite of these benefits, digital technology also brings risks to the education sector:

- ▶ As the number of mobile devices used by teachers and schoolchildren increases, there is a risk that the network infrastructure for the school/college will be overloaded, resulting in a negative impact on the bandwidth and performance of the technology used in the classroom.
- ▶ Security is another area of concern. There must be strict access controls in place to prevent students either intentionally or inadvertently accessing restricted areas on the school/college network, for example administration and finance areas. Network managers must implement controls and be able to monitor who is accessing the network.
- ▶ Students with access to technology can cheat when taking exams or carrying out controlled assignment work if they have (permissible) access to technology (of course not all exams will allow any digital device). This can be overcome by teachers having access to students' screens and monitoring what they are doing during live assessment.

Many education establishments (schools in particular) prohibit the use of mobile phones within the classroom to mitigate the risks of students continually accessing social media while they should be studying in lessons. Network managers will usually install firewalls and other software that will provide what is known as **application filtering** and prevent access to inappropriate and explicit content.

> **Key terms**
>
> *Application filtering*: this monitors applications being used and the types of data that comes from them, and blocks them if necessary. Network managers (such as those in schools and colleges) can dictate specific applications to be blocked automatically. This might include applications such as Facebook, WhatsApp, Twitter, etc.

As in other contexts, while digital technological has many benefits (and can be used to enhance and broaden students' learning), there are risks involved which require careful consideration. These risks must be considered carefully and mitigated against.

Transport

The transportation and logistics industries have also become more and more reliant on technology and although this brings many benefits, there are also many risks involved that need to be considered and mitigated against.

The increase in the use of technology means that these industries are more vulnerable to cyber-attacks, which can be costly and extremely disruptive. For example, as the transport sector expands, its use of technology to control and monitor its operations increases and so there is more data being transmitted between destinations, vehicles, ships, planes, trains, etc. These are a soft target, meaning they are vulnerable and easy prey for any professional hacker. In May 2021, a cyber-attack shut down the Colonial Pipeline which provides gasoline to around half of the east coast of the USA. This forced shutdown lasted around one week and the company has estimated that the attack cost them approximately $50 million. Similarly, a cyber-attack can put sensitive data at risk. A breach to sensitive customer, supplier or employee data can have serious legal, financial and reputational impacts on the business.

In addition, there is the risk to critical IT systems which can have major impacts. Consider critical IT systems that are associated with safety, for example a traffic light system that is controlled by AI. If there is a sudden failure, there is a potential risk of horrendous vehicle accidents. Another example would be aircraft that have many digital systems and, for example, make WiFi available onboard the aircraft, which links into the systems of airports and air traffic control, therefore leaving a digital footprint. This provides hackers with the opportunity to hack into the aircraft's operating systems or to interfere with communications and satellites. Failures associated with these systems could result in a risk to human life, damage to the environment and/or damage to property.

5.4.2 The potential impact of identified risks on the organisations and their stakeholders

Loss

- ▶ **Data** and information being stolen or corrupted so that it can no longer be used. If an organisation has their data and information backed up, then it can be restored in a day. However, whatever has been carried out since the last backup that can be used, will be lost.
- ▶ **Financial** – when data and information is lost or corrupted, a business can suffer financially. This can be due to the loss of orders and invoices, as well as the cost of 'person hours' for the lost work and for carrying out the restore and recovery process. If customers' personal data is compromised, then the organisation will legally be required to pay compensation. This can result in thousands of pounds being paid out and, in some instances, organisations have been known to go bankrupt for this reason. In addition, organisations have increased costs as they must improve the security of their systems and technology. Any cyber attack can create a situation where the customers lose confidence in the business and this can result in a loss of sales. Customers and suppliers may lose out financially as their information is obtained by the hackers and used for fraudulent activities. A customer may have their identity stolen and used to obtain loans and credit cards. They will have no idea this has been done until they get chased for payment.
- ▶ **Reputational** – when a business has been the target of a cyber attack, and data and information has been lost or stolen, there will invariably have a negative impact on the reputation of the business. The business will not be considered trustworthy by customers and/or suppliers and the stability of the company to function will be reduced. Customers may change to another supplier and suppliers may refuse to supply the business with the goods required to function.

Disruption

- ▶ **Operational** – when data has to be reinstalled due to a cyber attack, there is invariably an impact on the operation of the organisation. The installation of data and information can be time-consuming. In addition, organisations rely on data to carry out their business functions, whether this is internally or externally with customers and suppliers. Until the data is restored, the business is on hold.
- ▶ **Commercial** – the biggest impact for any organisation of a cyber attack is the impact on its day-to-day business activities. Depending on the type of attack, the business may not be able to function at all. The severity of the impact depends on the business sector. If a utility company were unable to function due to a cyber attack, then the consequences could be extremely severe, not only from a financial point of view to the business, but for their customers who could be without heat and light for long periods of time.

Test yourself

1. Explain the term 'chargeback'.
2. Discuss how a cyber attack can have an impact on an organisation.
3. Describe the three pillars of cyber risk management.
4. Identify four potential security breaches to an organisation's digital systems and technologies.
5. Explain what a digital download business is.

Skills practice

A UK company has started up a business to import cocoa products from the Côte d'Ivoire. The Côte d'Ivoire is a country situated in north-west Africa and it is well known for being one of the largest cocoa producers in the world. The business has links with the cocoa farmers who own the cocoa plantations. The business has commercial managers working in:

▶ Côte d'Ivoire
▶ Russia
▶ UK
▶ EU.

The business has an office and a warehouse. Within the office there is the director of the company, a finance department, marketing department, procurement department and administration staff. The warehouse has a warehouse manager and the employees who receive the cocoa products from the Côte d'Ivoire into the warehouse and despatch them to the locations identified by the commercial managers in each of the areas.

The business has a website which promotes the products, and an enquiry form for customers to complete to request information on the cocoa products and the prices.

Recently the director had a meeting with the commercial managers who travelled to the UK to discuss expanding the digital technology available to enhance the functionality of the business.

You work for an IT development company and have been asked to produce key information for the director:

▶ How the key areas within the organisation can use technology to improve the functionality of the business while meeting the needs of the stakeholders and customers, as well as ensuring the quality of the products/services offered.
▶ How digital technology can be used to market the cocoa products to a wider audience.
▶ How digital technology can be used for meetings between the director and the commercial managers without them having to travel to the UK or other destinations.
▶ The benefits and drawbacks of implementing digital technology.
▶ The risks, constraints and dependencies involved of implementing digital technology.
▶ An overview of how the implementation of digital technology can add value to the business.
▶ An overview of the risks involved of implementing a digital technology strategy.

Assessment practice

1 Explain the tertiary category of the business sector.
2 Discuss how digital technology and IT is used by the logistics functional area within a business.
3 Describe the commercial impact that a cyber attack can have on an organisation.
4 Identify four security breaches that are a risk to organisations when implementing digital technology.
5 Describe how digital technology has had an impact on the air travel industry.
6 Explain the difference between a product and a service.
7 Discuss the benefits and drawbacks of implementing cloud services.
8 Describe how a warehouse uses digital technology to enable automated stock/inventory control.
9 Identify three different ways in which rapid changes in technology and trends have had an impact on education.
10 Explain why the implementation of digital technology can create audience exclusion.

Content area 6: Data

You are going to learn about data and how it can be used to support business needs, as well as the range of ways in which it can be generated.

You will learn about the different forms of data and how organisations can use them to produce the outputs they require. Organisations can also choose between different structures in order to store data in files and folders.

Organisations use a range of data systems to help them collect, store, process, analyse and output data. You will explore the different functions of these data systems, as well as the business information tools used by organisations. You will also learn about the different data models that organisations can use to organise stored data.

All data needs to be managed, and the method will depend on how the data is collected, entered into the digital system and maintained. Most organisations need to analyse and process the stored data, using a range of data analysis tools, which you are also going to study.

After working through this content area, you should be able to apply an understanding of the use of data by organisations to support business needs. You will also have developed an understanding of the benefits and challenges that digital technologies present in terms of the creation and use of data.

Learning outcomes

In this content area you will learn about:

6.1   Data and information in organisations
6.2   Data formats
6.3   Data systems
6.4   Data management

6.1 Data and information in organisations

6.1.1 The differences and links between data, information and knowledge

Data is a big part of our lives. Lots of data is stored by governments, organisations, businesses and people. But anything that is spoken or written down can also be classed as data. If you break down any words, then you will find data.

Data is unprocessed raw facts and figures. Raw facts and figures means without meaning. It is not possible to understand what the letters and figures mean.

Two examples of data are:

NAT23%7

56GHS56

It is not possible to know what these mean. They are just a set of random letters, numbers and symbols.

The main points about data are:
- Data has no meaning.
- Data can be **qualitative** or **quantitative**.
- Data is raw facts and figures before it has been processed.
- Data can be made up of letters, numbers, symbols, graphics and sound.

Key terms

Data: raw facts and figures before they have been processed.

Qualitative data: non-numerical data.

Quantitative data: numerical data.

Information: data + [structure] + [context] + meaning.

Information is created when data is processed.

There is a formula for converting data to information. This is:

Information = data + [structure] + [context] + meaning

Sometimes it is not necessary to give data structure and context to become information, but it is always better to know the full formula.

The **structure** of data is how it is presented. One example of structure is for a date. One way a date could be structured is 24/04/2017.

The structure for this would be NN/NN/NNNN, where N shows that it should be a number.

Another example would be the structure for a postcode. The structure for a postcode would be LLNN NLL where N is a number and L is a letter.

The **context** of data is the environment that we know and understand to make sense of the data.

For example, data of:

15, cabbages, rabbits

means nothing. But when this data is put into a sentence, or context, the data makes sense.

15 cabbage plants were planted but the rabbits ate all of them.

The data now makes sense as a context has been provided.

The **meaning** of data is that it is in the correct structure and put into context.

How data and information are related

You have already learned about how information is:

data + [structure] + [context] + meaning

Data and information are related. This means that there are links between data and information. The main ones you need to be aware of are that:
- Information is in context while data has no context.
- Information is data which has been coded/structured.
- Data must be processed to become information.

Knowledge can be defined as:

'The ability to understand information and to form judgements and opinions and make predictions and decisions based on that understanding.'

(BBC, 2022)

The link between data, information and knowledge can be seen as:

Data → Information → Knowledge

Data is collected, which can then become information by applying structure, context and meaning, which in turn can be understood to become knowledge.

This relationship can also be represented by the DIKW pyramid shown in Figure 6.1.

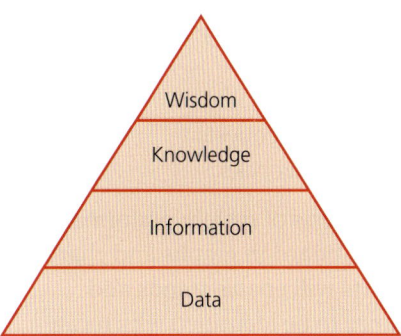

▲ Figure 6.1 The DIKW pyramid

The pyramid shows the relationship between data, information, **knowledge**, with **wisdom** at the top. As steps are taken upwards through the levels, questions can be asked and answers provided about the data. Each step upwards adds value to the data.

Data transforms into information by adding a meaning and/or context and/or structure. The linking of a range of data can also represent information.

As soon as information is processed, linked and stored, by a digital system or a person, it becomes knowledge.

Data represents information and knowledge at the fundamental, basic level.

Key terms

Knowledge: the ability to use information to, for example, form judgements and make decisions.

Wisdom: the ability to use knowledge to perform an action.

All businesses and organisations collect and store data. They will use this data to provide information which can then be used to make strategic decisions.

Activity

Select a business that provides goods to customers. Complete a DIKW pyramid for the business. Create a digital communication to present your findings to the owners of the business.

Test yourself

1 What are raw facts and figures?
2 How does data become information?
3 Identify two links between data and information.
4 Define the term 'knowledge'.
5 What does the W stand for in the DIKW pyramid?

6.1.2 Why organisations need data and information and how they are used

Data is an organisation's most valuable asset. It is said that an organisation that does not understand the importance of data is unlikely to succeed.

All organisations need data and information to be able to make informed and correct decisions. Without data and information, decisions cannot be made. An organisation will need to make decisions about a range of areas related to the function of the organisation.

Data is the raw facts and figures, including statistics, that an organisation will collect while carrying out its business function. Data needs to be processed and analysed to enable it to be put into a form which can be informatively used. Data can be qualitative or quantitative.

Organisations use data and information to enable them to:
- analyse market trends to identify patterns which inform decisions
- analyse system performance
- monitor users
- use targeted marketing
- inform strategic, tactical and operational decision making
- carry out threat/opportunity assessment (break-even, predictive models, cost analysis, market trends).

Analysing market trends to identify patterns which inform decisions

Market analysis is the assessment of the quantitative and qualitative characteristics of a specific market. The analysis of market trends can be carried out using data mining. Market trends can vary between sectors. For example, the holiday sector, traditionally, has an increased market trend in January and February. This is because, historically, people book summer holidays during these months.

The retail sector has an increased market share during October to December when gifts are bought for the various religious festivals such as Hannukkah, Divali and Christmas.

The fashion sector needs to predict trends and patterns in the clothes people will buy in advance. This is because clothes need to be made, shipped into distribution centres and delivered to the shops or put onto websites if the clothes retailer has an online presence.

Some trends can happen very quickly and can be influenced by social media.

An organisation that uses data and information to make decisions about market trends and patterns will be able to respond to these very quickly. This will help the organisation to stay competitive and profitable.

Analysing system performance

Data and information are usually stored on a digital system. It is highly probable that many of the employees will use a digital system to help them carry out their job roles. In addition, an organisation is likely to use a network to enable collaboration and sharing of data and information.

Over time the performance of a digital system can reduce. This may be as a result of, for example, an increased use of storage devices, conflict between software applications, or software updates.

Other reasons that system performance can reduce include outdated system components or an increased number of users.

Systems can be analysed using a range of software tools which can show where any reduction in performance is happening. Some tools can show hardware system **metrics**, including:
- central processing unit (CPU) and memory utilisation
- memory and socket interconnect bandwidth
- cycles per instruction
- cache miss rates

- type of instructions executed
- storage device access.

Other tools can show the software system metrics including:
- response times
- rate of completion of user requests
- identification of any bottlenecks.

The results from the analysis of system performance can be shown on a **dashboard** in graphs and charts. Using graphs and charts will enable trends and patterns in performance to be identified which can then lead to informed decisions relating to the possible upgrade of a digital system.

Key terms

Metrics: a set of numbers that gives information about a particular process or activity.

Dashboard: a type of graphical user interface providing simple visualisation of data related to performance indicators. It is commonly accessible by a web browser and can show real-time data updates.

Research

Investigate the software tools that can be used to provide metrics about hardware and software performance. Invite a system support technician to provide details about the tools used to monitor system performance in your centre or workplace. Discuss your findings with the technician and your teaching group.

Monitoring users

All organisations have users who use the digital system. These may be internal or external users.

Monitoring internal users was covered in Content area 4, section 4.1.6.

External users of the system will include those who visit the organisation's website. Website traffic monitoring software will provide an organisation with data and information. This information can be used to make informed decisions about any remedial action needed to increase the performance of the website or to enhance the design.

Monitoring users on a website can take many forms. These can include:

▶ checking how long a users spends on the website
▶ if the user is a registered user, how long they are logged in for
▶ how many clicks the user takes to carry out the purpose of their visit.

This data can be used to improve the user experience: if the website is intuitive and easy to use, it is likely that users will visit the website and be able to complete their intended task quickly.

> Data mining is covered in section 6.4.2 of this content area.

> Networks are covered in Content area 7, section 7.2.

> Hardware components are covered in Content area 7, section 7.1.2.

Using targeted marketing

Having data and information about their customers will enable an organisation to carry out targeted marketing. This data and information can come from a range of sources. Data may be analysed about specific products bought on a website by a customer which can lead to an organisation targeting marketing about 'similar but different' products.

For example, if a customer regularly purchased dog food from an organisation, then the data would show that there was a high probability that the customer owned a dog. The organisation could then send this customer targeted marketing about other dog products. The targeted marketing could be, for example, an email, or recommendations next time a customer visited a website.

Informing strategic, tactical and operational decision making

Organisations can use data and information to help them make strategic, tactical and operational decisions. These three types of decisions are linked and depend on the results of the analysis of data and information.

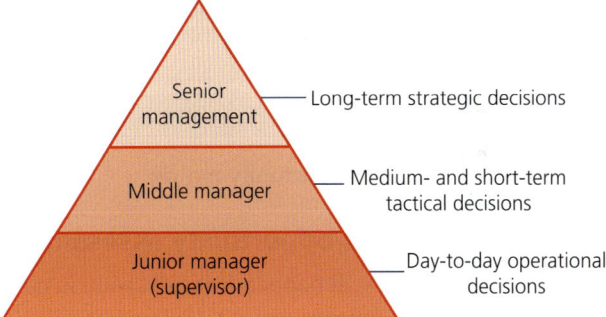

▲ Figure 6.2 The staff in an organisation use data and information to help them make strategic, tactical and operational decisions

Table 6.1 shows the differences between the three types of decisions.

	Strategic	Tactical	Operational
Time	Long term	Medium term	Short term day to day
Type	Complex		Simple, routine, repetitive
Level	Senior managers/directors	Middle managers/heads of departments	Junior managers/supervisors
Example	Negotiating new contracts with suppliers	Increasing or decreasing number of employees	Creating staff rotas

▲ Table 6.1 The differences between strategic, tactical and operational decisions

Further details about strategic planning and decision making can be found in Content area 3, section 3.1.2.

To help inform decisions at strategic and tactical levels, data and information can be used to carry out a threat/opportunity assessment. A matrix like the one shown in Figure 6.2 is used to analyse the threats and opportunities of implementing a decision.

	Threats	Opportunities
Short term		
Long term		

▲ Figure 6.3 A threat/opportunity matrix

Carrying out threat/opportunity assessment

Threats refer to what will happen if the decision is not implemented, while opportunities refer to what will happen if the decision is implemented.

A threat/opportunity matrix (see Figure 6.2) can be used by an organisation in a range of strategic and tactical decision-making processes. These include:

▶ Break-even and cost analysis: The break-even is when the costs of, for example, producing a product match the income generated from that product. Cost analysis relates to the benefits of a decision measured against the cost of implementing the decision. For example, the cost of implementing cloud-based storage against the purchase of an on-site server, technical staffing and maintenance costs.

▶ Predictive models, including market trends: Data can be used to predict what might happen based on previous data. For example, on Valentine's Day the sale of flowers and chocolates generally increases. Based on the previous years sales, a business selling chocolates can ensure that they have enough stock of the most popular chocolates. This can be achieved through analysing data and looking at trends in sales.

Analysing market trends to identify patterns which inform decisions was covered earlier in this section.

Activity

Complete a threat and opportunity matrix for:

▶ an online retailer stocking the latest emerging trend in children's toys
▶ building a new teaching room in a college.

Discuss your matrix with the rest of your teaching group. What are the similarities and differences?

Test yourself

1 Identify two uses of data by an organisation.
2 Identify one hardware and one software metric.
3 Identify one way that an organisation can collect customer data.
4 At what level are operational decisions taken?
5 What is meant by 'an opportunity'?

6.1.3 How data is generated

Industry tip

Every organisation, irrespective of its size, generates, stores, processes and analyses data to help it flourish. What the data is will depend on the organisation sector. Irrelevant data can be worse than having no data so all data generated, stored and processed, and how it is analysed, must be relevant to the organisation, its function, needs and requirements.

Before data can be used, it has to be generated. Data can be generated in a number of ways including:

▶ human generated
▶ AI/ML
▶ sensors
▶ IoT
▶ transactional data.

ML is covered in Content area 3, section 3.2.1.

Sensors are covered in Content area 7, section 7.1.2.

IoT is covered in Content area 3, section 3.2.1.

Human generated

Human generated data is just that. It is the data that is generated every day by people using digital systems. Examples of human generated data include:

▶ emails
▶ social media posts

- digital communications and files, for example documents, spreadsheets and presentations
- images, audio and video files.

The Conversation estimated in 2021 that in any given day:

- 500 million tweets are sent
- 294 billion emails are sent.

The Conversation (2021)

Research

Investigate the type and amount of data generated by humans over a year. Create an infographic to show your findings.

AI and ML

AI is the science of training machines to perform human tasks. This includes speech interfaces that are used in, for example, smart devices that many people have in their homes.

ML is a subset of AI. It teaches a digital device how to learn. ML looks for patterns in data and uses these patterns to attempt to create conclusions just as a person would. For example, it is raining outside so I will need a coat and umbrella.

ML uses examples from the real world and learns what to do by analysing those examples. This is done by using an algorithm. When the algorithm has been perfected, as far as it can be, this knowledge is then applied to a new set of data. How an ML machine learns can be seen in Figure 6.3 which shows the life cycle of ML.

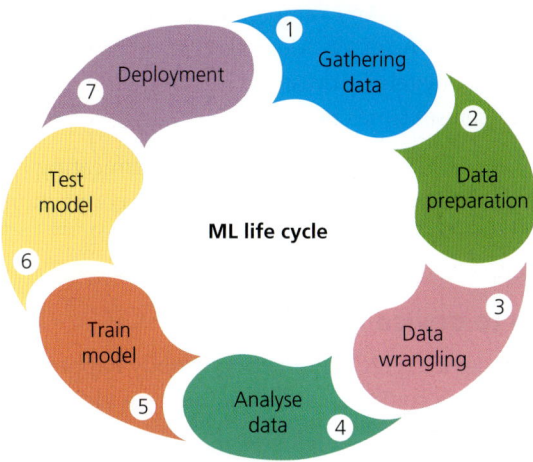

▲ Figure 6.4 The ML life cycle

Both AI and ML need large amounts of data to be able to carry out their processes. They also need granular data which is diverse in meaning. This will enable patterns to be found and so begin the learning process.

Research

Investigate how a driverless (autonomous) car learns. Consider the initial data and the data that is used when the car is driving on the roads. Think about future developments in autonomous cars and any limitations that will need to be overcome. Discuss your findings with the rest of your teaching group.

Sensors

Sensors can carry out a range of processes which can result in data being generated. Sensors can be used in a range of applications. A sensor will measure a physical item and covert this into a digital signal. This signal is used to provide inputs to a digital system to provide data. For example, cars have lots of sensors, such as a parking sensor that can produce an audible output when the car is within a specified range of an object.

Sensors gather data using a range of methods. For example, a sensor can be used at the entrance of a café, to gather data related to how many customers enter the café and at what times. This could help, for example, to ensure that the optimum number of employees are available at busy times and can also help with stock control.

There are a range of physical items that can be measured using a sensor. Temperature in a greenhouse can be measured. If the temperature goes above a predefined level then a digital system can increase ventilation. Sensors can be used to control robotic packing areas. These sensors keep the robotic machines on predefined routes.

Food storage can also use sensors. These work in the same basic way as a sensor in a greenhouse: the sensor detects any fluctuations in temperature and sends this data to a digital device. The digital system will then make any amendments to the temperature to ensure the food stays in peak condition. It is possible that an alert can be sent to a person, but this is more than likely to be advisory as the sensor and digital system will have rectified any issues identified. The organisation can use this data to identify trends and patterns and take any remedial action needed.

The IoT

The IoT is a collection of smart devices which can connect to each other and exchange data using the internet. An increasing number of people have at least one smart device in their home. This may be a smart speaker, smart lights or smart heating.

IoT devices typically include sensors which track and record data. Associated devices and technology – that is ones that are connected via the IoT – can monitor and measure that data in real time. The data is also stored and can be retrieved at any time.

IoT devices connected to the internet means the devices can:

- ▶ collect data and send it through the internet
- ▶ receive information from the internet
- ▶ or both.

All smart devices have four different components:

- ▶ sensors
- ▶ connectivity
- ▶ data processing
- ▶ a UI.

The sensors collect the data and transmit it over the internet. The data can be stored to be used at a later date, for immediate processing or to be sent somewhere else as required.

For example, recommendations about new products can be made based on previous shopping history. Other examples include the recalculation of a journey based on the information provided about hold-ups on the current route. Data can also be collected, stored, processed and shared in a **Smart City**.

Data collected by IoT devices can be processed into more useful information for decision making. The world is flooded with data and one of the main functions of the IoT is to gather huge amounts of data to improve operations and functionality. Individual data collected could, for example, be hours of footage from a surveillance camera, how much exercise has been completed in a day or viewing figures for a TV programme.

These are examples of individual data but imagine the vast amounts of data that can be collected, processed and analysed from organisations on a global scale. Just consider social media platforms such as Facebook, Twitter, WhatsApp and LinkedIn.

Device to device

Device to device (D2D), also known as machine to machine (M2M), connections occur when data is transferred from one device or 'thing' to another over a network. Devices include sensors, robots, computers and mobile devices. An example of D2D is a home assistant such as Amazon Alexa or Google Nest automatically switching on smart lighting light bulbs based on user-set times before the user arrives home.

Another example would be a laptop communicating with a printer using a Bluetooth or wireless connection. This communication is because they are directly connected.

Human to device

Human to device (H2D) interaction usually happens using a UI. This could be carried out using a visual

UI or with a voice command to a digital assistant such as Amazon Alexa. The human will speak to the digital assistant using command words that the device 'understands'.

Another example is interacting with a heating or lighting control UI to provide user-set values. For example, setting the bathroom heating to come on at 6.30 a.m. and off at 8.30 a.m. on a weekday.

Transactional data

Transactional data is data about transactions for an organisation. Transactional data can be created between organisation and organisation, or organisation and customer.

For example, you buy and pay for an item from an online retailer. Your name and address, the item(s) you bought, the details about the item(s) such as stock number, colour and size, payment method and delivery are all examples of transactional data. The transactional data will also be used to amend stock levels of the item(s) that was bought.

Organisations can use transactional data to identify trends and patterns that are occurring in any interactions the organisations may have. Stock levels and production rates could be amended as a result of analysing transactional data. It is also possible that transactional data can be used for forecasting future cash flows which may lead to a change in process. This may include the forecasting of the processing of any raw materials that are used during manufacturing or cost processing of any items the organisation may sell onto a third party including customers.

Transactional data may also be used to produce targeted marketing to customers.

Organisations also generate transactional data internally. For example, payroll is an example of internally generated transactional data.

What organisations do with the data generated will depend on the sector: the data may be collected to inform future decisions or as part of a research and development (R&D) plan.

6.2 Data formats

6.2.1 The forms that data can take and their implications for use and analysis

Data can take many forms. To be of use it is important that data is stored and used in a form that can allow the user to successfully manipulate it.

It is, therefore, very important to consider the way in which data is stored before it is processed.

Data types

There are many different data types. Which data type is selected will depend on what data is to be stored and how it will be used and/or processed.

The main data types are:
▶ Boolean
▶ character
▶ date
▶ integer
▶ real
▶ string.

Data types are covered in Content area 2, section 2.1.1.

Table 6.2 shows the data types and the relevant Python constructs.

Data type	Description	Example of data	Python construct
Boolean	There are only two choices, e.g. true or false	Yes or no True or false 1 or 0	bool()
Character	Stores a single character which can be a letter, number or symbol	$	Not used in Python as this would be a string with a length of 1
Date	A date with a defined structure. Which format is used will depend on how the date is to be stored and processed	12/02/2025	datetime
Integer	Whole numbers which can be positive or negative	1960 −97	int()
Real	Any number, with or without decimal places. These can be positive or negative	12.30 −4763	float()
String	Stores alphanumeric combinations and text. A group of characters stored together as one	3T6HjV Oxford Place	str()

▲ Table 6.2 Data types and their Python constructs

Research

Select one key business area in your centre or workplace. Investigate what data is stored, the data type and how the data is processed and used by the end users for that business area. Present your findings to your group.

If the data is stored as an incorrect data type, then any processing or analysis needed to be carried out on the data may be problematic. This may result in incorrect processing and analysis which could, in turn, lead to incorrect decisions being made by the data holder.

Data can be shared. For example, the results of research from one university can be shared with others researching the same area. Governments share data either within the country or with other countries across the world.

Businesses can also share data. For example, a business may advertise on a commercial TV station. The results of the analysis of the impact of the advertising will be shared with the business by either the TV company or the company that analysed the data. This could help inform strategic decisions relating to, for example, advertising budgets, new products or target market.

Data format

For data to be shared without being corrupted, common file formats should be used. A common file format will enable the contents of the file to be stored, processed and used across a range of software platforms.

The main file formats are:
▶ American Standard Code for Information Interchange (ASCII)
▶ comma-separated values (CSV)
▶ fixed-width text file
▶ JavaScript Object Notation (JSON)
▶ Extensible Markup Language (XML).

ASCII

ASCII is a 7-bit character set containing 128 characters. It contains the numbers from 0 to 9, the upper- and lowercase English letters from A to Z, and some special characters.

The character sets used in digital devices, in HTML and on the internet, are all based on ASCII. ASCII codes enable the storage devices and screens to communicate to enable people to use digital devices. ASCII is used to ensure that all digital devices use the same format which enables, for example, the sharing of documents and files. This means that a device using Windows is able to, for example, share a word-processed document with a MAC.

ASCII is used to translate computer text to human text and vice versa. When keys are pressed on a keyboard each letter is sent as a group of numbers. As the range of standard ASCII is 0 to 127, it only requires 7 bits or 1 byte of data.

Upper- and lowercase characters are assigned different numbers. For example, the character S is assigned the decimal number 83, while s is assigned decimal 115.

ASCII is a shared means of communication between computers.

CSV

CSV is a file format where data is arranged in columns where the columns are delimited (separated), by a comma. Each line will represent a complete record, with tabular data stored as plain text. All records should have the same number of fields, in the same order.

CSV is a common data file format that is often used by business to move data between software that may be incompatible. Many data-handing software packages support the saving, importing and processing of CSV files.

One way a CSV format can be used is when transferring data between two applications which are in different formats, for example a spreadsheet and a database. The CSV format can be used to enable the data to be saved and opened in both applications.

Fixed-width

Fixed-width is a file format where data is arranged in columns, but instead of those columns being delimited by a certain character (as they are in CSV) every row is the exact same length.

The format is specified by column widths, pad character and left/right alignment. Column widths are measured in units of characters where each column has a defined width.

A fixed-width text file must follow these rules:
- Each row contains one complete record of information.
- Each row contains one or many pieces of data: columns/fields.
- Each data column has a defined width specified as a number of characters that is always the same for all rows.
- The data within each column is padded with spaces or another character such as a / or !. These are used if the data does not completely fill all the characters allotted to it.
- Each piece of data can be left or right aligned, meaning the pad characters can occur on either side.
- Each column must consistently use the same number of characters, same pad character and same alignment (left/right).

(Softinterface, 2022)

For example, customer data could be saved in a fixed-width file format. Part of the customer data is shown in Table 6.3.

Name	Country
John Smith	UK
Daisy Patel	Germany
Sundip Rawstone	Russia

▲ Table 6.3 Some customer data in fixed-width format

In this case:

- Each record is 30 characters long.
- The first column is 20 characters. It is left aligned. It is padded with spaces. It contains the NAME data within it.
- The second column is 10 characters. It is left aligned. It is padded with spaces. It contains the Country data within it.

A fixed-width font, such as Courier, should be used to best display the data held in a fixed-width file. The file extension for fixed-width files is usually .txt.

> ### Activity
>
> Find a data set on a website that relates to collected statistics for your local area. Create a fixed-width file from the statistics.

CSV and fixed-width files can be used to import data into, for example, a spreadsheet or database. This will enable analysis to be carried out on the data. Which format the initial data file is in will be shown by the file extension used.

JSON

JSON is a text-based file format that is often used in web applications. The format represents structured data and is based on the JavaScript object syntax. The data being exchanged can only be text.

JavaScript objects are converted into JSON, a text-based format, which is then sent to the server. The reverse process then happens with the server sending JSON which is then converted to JavaScript objects to be displayed on the webpage.

This means that webpages can send and receive data with limited issues and problems occurring which will enhance the user experience.

Many different programming languages include standard code to generate and **parse** JSON-format data. The file extension for JSON files is .json.

> ### Key term
>
> **Parse:** the formal analysis by a computer of a sentence/string of words into its constituent parts.

XML

XML is a file format that stores and transports data. XML is a W3C recommendation. It is a World Wide Web mark-up format like HTML which can be used as an input format. XML was created to ensure that data can be read by devices and people. However, this file format does not carry any information about how the data is to be displayed. XML stores data in plain text format. This means that an XML file format, as with the CSV and fixed-width file formats, provides a software/hardware-independent way of storing, analysing and sharing data.

Data can be input in an XML format and can then be converted to a different format to enable the data to be processed and analysed, for example by a spreadsheet or database. The same XML file can be used to present the contents in many different applications.

A key benefit of using an XML file format, as with CSV and fixed-width file formats, is that the file format can simplify and enhance the availability and use of data.

XML and spreadsheets are both able to process structured data. Most XML files are converted to a spreadsheet format. However, there are some examples of text being input using an XML format and converted to a document. Some organisations use Microsoft InfoPath for designing, distributing, filling and submitting electronic forms containing structured data/information in an XML format.

Not using shared or compatible formats for data can be costly and complicated for businesses. When incompatible formats are used, any processes that involve data – for example, exchange between systems or systems upgrades – will need to involve translation or conversion of that data into the relevant format. Using CSV, fixed-width and XML file formats can limit these impacts and ensure that full and complete processing and analysis of the data can be completed.

> Further information about W3C can be found in Content area 4, section 4.2.4.

6.2.2 The difference between file-based and directory-based structures, and how they are used in data analysis

File-based structures

A file-based structure is used to maintain and organise single or many data files and can help with basic data management. A file-based structure facilitates a range of application software packages to carry out functions for end users of the digital system. Each package defines and manages its own data. This can put limits on how the data can be used or transmitted. It is important that the system permits concurrent access by different processes. Data stored in a file-based system should be consistently structured and stored so it is accessible.

Historically, each area within an organisation entered, saved, processed and analysed its own data.

> Key areas of an organisation were covered in Content area 5, section 5.1.2.

Each area within an organisation dealing with its own data causes issues. These can include:
▶ different file formats being used which can lead to incompatible data between areas
▶ data duplication
▶ lack of flexibility in organising and querying the data
▶ increased number of different application programs.

Directory-based structures

A directory-based structure is the way that files and folders are displayed to the user. This is often in the format of a hierarchical tree structure.

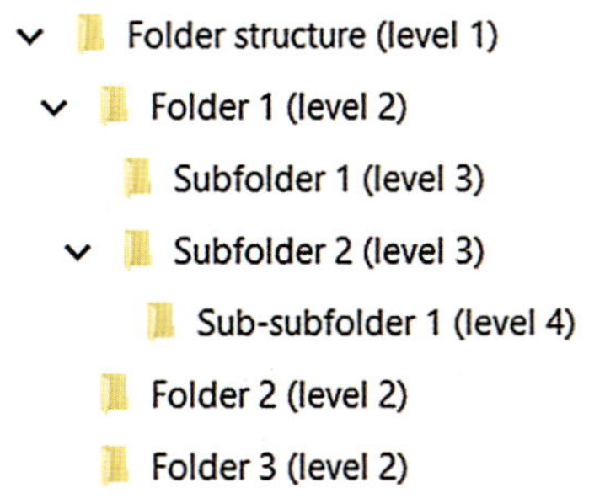

∨ 📁 Folder structure (level 1)
  ∨ 📁 Folder 1 (level 2)
    📁 Subfolder 1 (level 3)
    ∨ 📁 Subfolder 2 (level 3)
      📁 Sub-subfolder 1 (level 4)
    📁 Folder 2 (level 2)
    📁 Folder 3 (level 2)

▲ Figure 6.5 A folder structure

A directory structure can make it easier for the end user to locate files as the folder structure should be logical.

For example, an HR department may store records of all employees. Each employee will work in a department. Each department could have its own folder, with subfolders for each employee. Within the employee's subfolder will be stored all files relevant to that employee.

6.3 Data systems

6.3.1 The features and functions of data systems and their importance to organisations

All organisations use data and data systems to be able to store, process and analyse data. Some of this data will be generated internally within the organisation while other data will be taken from other sources.

There are different features and functions of data systems which enable organisations to productively process and use data.

Data wrangling

Data wrangling is the process of changing unorganised and raw data into standardised data that can be analysed, so making it useful. Data that is the output of successful data wrangling means that the output data is structured to meet the needs of the organisation. There is so much data available, for example in big data stores. However, much of this data is in an unstructured and raw form. To make this data useful an organisation would carry out the process of data wrangling.

The main steps in the data wrangling process, which take an **iterative** approach, are shown in Figure 6.6.

There are six main tasks in the data wrangling process.

1 Discovery

The initial step is to understand the data and fully understand what the data is about.

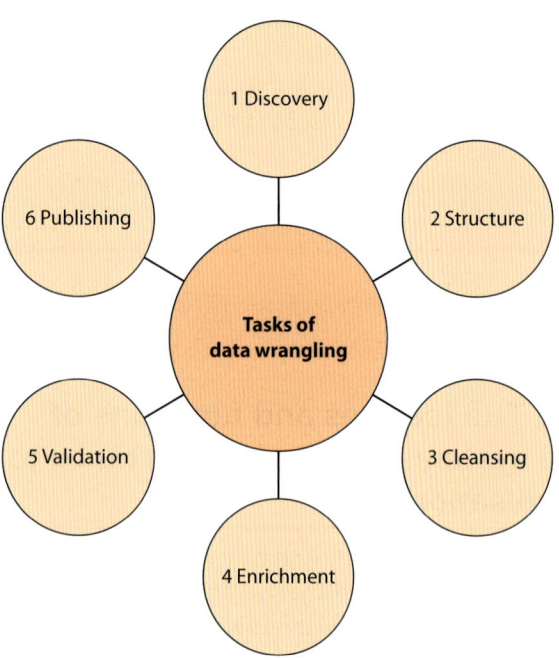

▲ Figure 6.6 The data wrangling steps

Understanding the data will enable the best outcomes for its analysis. For example, a big data set about people could be used to analyse the age demographics across different counties of England.

2 Structure

Most data, including big data sets, can be in an unstructured and disorganised form. This means there will be no structure to the data.

> The structure of data was covered in this content area, section 6.1.1.

Therefore, in the next step, data will need to be structured to ensure that it is accessible. The structure that is used in this step will depend on the requirements of the output

of the analysis. For example, the original data set may have address and postcode in the same field. It may be that the postcode needs to be in a separate field to enable successful analysis of the data based on the specified requirements.

3 Cleansing

Every data set includes data that does not fully meet requirements and so it can skew the final analysis results. The third data wrangling task is to cleanse the data set to ensure that any anomalies are removed. For example, some data may include special characters, or repeated values. It may also be that different formatting has been used during the collection of the data sets. This cleansing step attempts to remove all the anomalies in the formatting of the data.

For example, if the data relates to the population of the UK, it may be that the four nations have been recorded and formatted differently. Northern Ireland could have been recorded as NI, N.I or Nor Ire. By cleansing the data these can be replaced with a single data entry, for example N.I. By doing this the analysis being carried out on the data set will provide full and complete results which can then be processed and used.

4 Enrichment

It is unusual that any externally collected data set will include all the data requirements to meet the specified needs. It may be that two, or more, data sets will need to be combined, or data will need to be enriched. The task of enriching data includes adding to or combining data sets so that the results of the processing and analysis meet the specified requirements fully and completely. For example, adding data relating to the number of children born in a specified year will enable an analysis of the future need for pre-school places to be carried out.

5 Validation

Before the data set can be analysed, the penultimate data wrangling step is that of validation. Validation, in the data wrangling process, means that the reliability, quality and safety of the data is authenticated. This involves checking the data to ensure it is complete and all data in a given field meets, for example, the same structure and level of completeness.

6 Publishing

The final data wrangling step shown in Figure 6.6 is that of publishing the data. This is when all data wrangling steps have been completed on the data. The data is full and complete and will enable analysis to be carried out producing a full and complete output.

Activity

Create a questionnaire to collect names, birth dates, course names and subjects from 15 people in your centre. Do not provide any information relating to the format of the answers.

Share your data with three members of your teaching group. Carry out data wrangling on the 60 records you have collected.

Document your findings for each step. Does your final wrangled data match that of the rest of the people you shared your initial data with?

Core functions

All data systems have the same basic functions. How an organisation uses these functions will depend on the data being processed and analysed, and how the results are to be used.

The core functions of data systems are:

▶ **Input** – is the collection of raw data.
▶ **Search** – searches can be carried out on the data to meet the specified needs and requirements of the organisation. Data that has been saved in the data system can be searched many times using different search criteria.
▶ **Save** – data that has been input into the data system can be saved or stored so it can be used again. Saved data can be searched, processed and analysed. The saved data can also be edited and resaved.
▶ **Integrate** – different data types and formats can be integrated into a single location. This produces full and complete outputs that meet specified needs and requirements.

> Data warehouse and warehousing is covered in this content area in section 6.4.2.

▶ **Organise (index)** – saved data can be organised, indexed, to ensure that it meets the needs and requirements of the end users. An index can be used to increase the speed of searching data. Indexes enable the searching of a data system without having to search, for example, every record in a database. Indexes can be efficient and time-saving.
▶ **Output** – the processed and analysed data, the output, are sent to the relevant people or places.
▶ **Feedback loop** – feedback is output that is returned to, usually, senior management, to help evaluate the process to, for example, correct the tasks carried out at the input stage.

Data entry and maintenance

Data has to be input into a digital system. Sometimes this is done by combining data stores, but data is usually initially input by a person. Every person is capable of making errors during data entry. There are, however, features that can be introduced to reduce the number of data entry errors.

There are two main types of errors that can be made on data entry when taking data from a paper-based source document or when a user is entering data on an online form or survey. These are:

▶ transcription errors
▶ transposition errors.

Transcription errors occur either when copying the data from the source document for data entry or by a user entering data online. For example, this type of error can occur by pressing the wrong key on the keyboard or hitting two keys at once. Examples of transcription errors are shown in Table 6.4.

	Incorrect	Correct
Postcode	TN18 7TH	TN28 7TH
Surname	Stuagt	Stuart
Date	25th March 2004	26th March 2004

▲ Table 6.4 Examples of transcription errors

A transposition error occurs when two letters or numbers have been reversed. Examples of transposition errors are shown in Table 6.5.

	Incorrect	Correct
Postcode	TN82 7TH	TN28 7TH
Surname	Staurt	Stuart
Date	26th March 2040	26th March 2004

▲ Table 6.5 Examples of transposition errors

Using **validation** and **verification** techniques can reduce data entry errors.

Key terms

Validation: checks that the data being entered into a digital system is sensible and reasonable, and checks it against pre-set rules.

Verification: a check to see whether the data being entered into a digital system is identical to the source document or initial data entry.

Validation

Validation is a check that is run by the digital system as the data is being entered. Validation attempts to prevent the entry of any data that does not meet the predefined rules. The rules will not stop incorrect data being entered but ensure that the data being entered is:

- sensible
- reasonable
- within predefined boundaries
- complete.

> Validation techniques and how to write (code) them in Python were covered in Content area 2, section 2.6.1.

Validation techniques can be used on an online data-entry screen to limit the risk of data entry errors. If an error is made, then a useful message should be provided to the user. The message should provide details about the error and how it can be corrected. It is also common to use colours to help the user. For example, the use of green when everything has been entered correctly and red if an error has been made.

Data entry errors can result in incorrect data being held. This is an example of **GIGO**. This means that as incorrect data is entered, the data stored will also be incorrect. When any processing of this data occurs, the results of the processing will be correct in terms of the data that was processed but will be incorrect in terms of accuracy and usefulness.

Key term

GIGO: Garbage In, Garbage Out.

Activity

Find a website that includes a data entry screen, for example a retail website. Investigate the validation techniques that have been used and the error messages that are shown when a data entry error has been made.

Discuss with your group how using validation techniques and error messages will decrease the impact of data entry errors on the owner of the website.

Most data in industry will be entered through a data-entry screen. When a data-entry screen is being designed and created, it is important that the developer takes the time to clearly understand the needs and requirements. For example, where data needs to be in a specified format, this must be included in the validation rules set for each of the data-entry fields. By doing this, the time taken to enter the data will decrease as validation techniques, such as presence and type checks, will help users enter the data and also increase the probability that the correct data will be entered first time.

When the data has been entered, and during the life of the data in terms of its usefulness, it will need to be maintained.

Data can be maintained in various ways, for example, by carrying out regular scheduled searches to remove redundant or expired data. Legislation also requires those storing data to delete data when requested to do so by the data owner.

Research

Research and make notes about situations when data should be maintained and the risks of not maintain the data. Discuss your findings with the rest of your group.

> The different data types were covered earlier in this content area.

Visualisation

When data has been input, processed and analysed it has to be output, that is presented, in a format that is useful to the end users. The main ways data can be presented are:

- graphs/charts
- data tables
- reports
- infographics.

Graphs and charts

Graphs and charts are generally used to present numbers. A graph/chart will enable the end user to visualise and understand the data more easily than being presented with just numbers. Titles and labels can be used to put the data being shown into context. It is easy to identify trends and patterns using a graph or chart.

There are some disadvantages to using graphs and charts. If the graph or chart being used to present the numbers is poorly presented, the end users can misinterpret the data being shown.

There are many different types of graphs and charts. Each type has a different purpose and this should be considered when selecting the type of graph or chart to use to present the data.

The most commonly used types of graphs and charts used to visualise data are:

- area chart
- bar graph/histogram
- column chart
- dual axis chart
- line graph
- pie chart
- scatter plot
- stacked bar graph.

> ### Research
>
> Research each type of chart in the list, including the type of data that each can be used for. Create a communication to present your findings for an audience aged 16 to 19 years.

Data tables

Data tables can be useful when the data being presented belongs to the same category, or when a single category of data varies when measured at different points or time. For example, the percentage increase in the use of the train network for different regions of England.

To successfully use a data table, the data being presented and visualised must be relatively small. It is very difficult for an end user to interpret or visualise large amounts of data in a table, which is simple in structure. It can be difficult to represent precise data on a graph or chart, but a data table can allow the clear presentation of numbers to two, or more, decimal places.

Data tables can be used to visualise data in a clear, easy-to-read and understandable format. They can be used to summarise data. However, the data tables must be clearly labelled, with headings used to indicate what the table is showing. Column headings must also be clear to enable an end user to fully understand the data being presented and visualised.

Reports

Data can be presented in data reports. There are many different types of data reports but in industry most reports will be formal and will present an overview. The data being presented and visualised must be the most relevant, with no errors. Errors could include selection of the incorrect data or incorrect formatting. When creating the data report, essential information should be extracted from the data so that the data report conveys all the data needed to fulfil the specified requirements and needs of the end user. The data presented in the report must be arranged and displayed in an easy-to-read format to enable the end users to easily visualise and understand the data report.

Reports can be created in word processing software which will enable a range of different components to be included. These elements could include graphics, images, extracts from a database or spreadsheet and graphs / charts. It is likely that an organisation will have a template which incorporates the house style. Using a pre-defined template will probably negate the issue of important information and data being omitted.

Reports can be provided to the target audience in hard copy or online. If the reports are to be presented online by, for example, an attachment to an email, or hosted on an internal or external website, the range of elements could be increased to include hyperlinks to, for example, dynamic charts taken from a spreadsheet. Contents pages can also include internal document hyperlinks which will enable a user to 'jump' to the section they need to focus on.

Reports can also be elicited from financial modelling, spreadsheet and database software. These types of software have built-in functions that enable data and information to be queried and sorted. Calculations can also be included which, for example, will enable sales totals and percentage profit to be included. This will enable reports to be created which will fully meet the defined requirements, for example sales figures for a cloud service provider based on the size of the client organisation. The reports can be formatted to meet the needs of the target audience. Again, it is probable that the organisation's house style will be incorporated into these reports.

> ### Research
>
> There are many different types of reports that can be used in the digital industry. Research the different types, including when and why each type would be used. Make notes about your findings.

Infographics

According to the *Oxford English Dictionary*, infographics are,

> 'a visual representation of information or data'.

An infographic is a collection of images, charts/graphs, with minimal text to provide an easy-to-visualise overview of a topic.

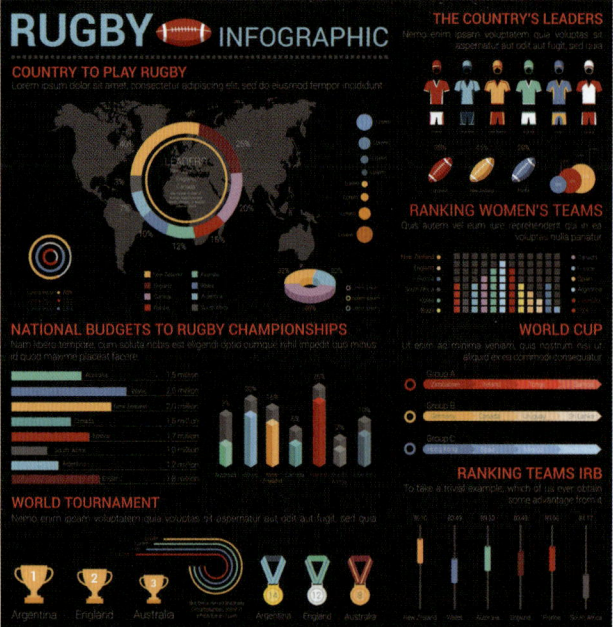

▲ Figure 6.7 An infographic

An infographic provides information in a visual and accessible format, often with minimal use of text and using charts or graphs. Looking at data can become difficult for the end user to visualise the message of the data. By displaying data in a visual way, the end user can visualise and increase their understanding of the data so increasing effectiveness.

Research

Research data visualisation infographics, finding examples. Discuss the effectiveness of each example with your teaching group.

Activity

Create an infographic to show the different methods that can be used to visualise data.

Test yourself

1 What are the steps included in the process of data wrangling?
2 What is a transposition error?
3 What is the purpose of a format check?
4 Identify two different charts and graphs.
5 What is an infographic?

6.3.2 The purpose of business information tools and their use in business

Business information tools are software which aim to provide data for, and support the making of, informed decision making. The main purpose of these tools is to gather, organise, process and analyse data to show trends and patterns.

Business intelligence software

The purpose of business intelligence software (BIS) is to gather, analyse, process and report data to provide business intelligence. BIS usually gathers data from a data warehouse. Data from BIS can be displayed on a dashboard. Used correctly, a spreadsheet is a basic type of BIS.

An organisation that does not value data is unlikely to succeed. BIS enables managers at strategic and tactical levels to understand the results of data processing and analysis to make informed decisions. BIS can provide processed data that will enable trends and patterns to be highlighted.

Financial planning and analysis software

The purpose of financial planning and analysis software is to support activities including planning and setting budgets, and budget forecasting and modelling.

The software will enable a business to create a financial model to test a range of financial scenarios. For example, the model could be changed to predict the impact of any possible price rises and how this could affect the business.

There are advantages to using this type of software. For example, models can be created and shared between employees, and different numbers can be locked so they cannot be changed, this would most probably be related to fixed costs.

One disadvantage of using this type of software is that while planning and analysing the financial aspect of a business is beneficial, it can be difficult to predict the impact of world events on business costs.

Using financial planning and analysis software enables a business to create and forecast the financial aspects to make informed decisions. These decisions are likely to be taken at strategic level.

Data warehouses and data mining are covered in section 6.4.2 of this content area.

APIs are covered in section 6.4.5 of this content area and Content area 8, sections 8.1.3 and 8.2.3.

Research

Investigate the different financial planning and analysis software. Identify the features each software includes. Make notes about your findings.

Customer relationship management software

The purpose of customer relationship management (CRM) software is to make customer management less time consuming and easier for those employees who are customer facing. A CRM software package helps to track, manage and record customer interaction. CRM software can gather data from a range of sources including the organisation's website, social media and telephone, email, and live-chat interactions with customers. A CRM system can also utilise an API.

CRM software can be used by an organisation to analyse the gathered customer data from a range of sources. Techniques used when analysing the data gathered by a CRM system include data mining and the identification of trends and patterns. The results of the analysis can be presented in a visual way, including the use of charts and graphs. By presenting the analysed data in these ways, informed decisions can be made at strategic and tactical levels.

Strategic and tactical decisions were covered in section 6.1.2 of this content area.

Charts and graphs were covered in section 6.3.1 of this content area.

6.3.3 The features of different data models and how organisations use them to organise data

One definition of a data model is:

'The logical interrelationships and data flow between different data elements involved in the information world that also documents the way data is stored and retrieved.'

(ScienceLogic, 2022)

A data model will help when a database is being designed and will ensure that the final database is fit for purpose and has no omissions. The final database will be efficient, meaning that all database relational tables, primary and foreign keys are fully and completely defined, which will lead to little chance of any omission or errors in the structure.

This will ensure that all the data needed by the database, based on the requirements of the end user, will be correctly and fully represented, with no missing or redundant data.

Key term

DBMS: database management system.

Type	Conceptual data model	Logical data model	Physical data model	Hierarchical database model	Relational data model
Definition	Defines what the system contains.	Defines how the system should be implemented, regardless of the database management system (DBMS).	Describes how the system will be implemented using a specific DBMS.	Shows the database structure on a hierarchical (tree) structure.	Shows a database as a collection of relationships.
How used to organise data	To scope, organise and define rules, and to identify the data used.	To develop the rules and data structures. This model expands on the conceptual modelling.	The actual implementation of the database.	To show the relationships between the tables, records and fields.	To show the relationships between the tables, records and fields using primary and foreign keys.
Techniques	ERD	ERD Data Dictionary	ERD	ERD	ERD Data dictionary

▲ Table 6.6 Types of data model and their uses

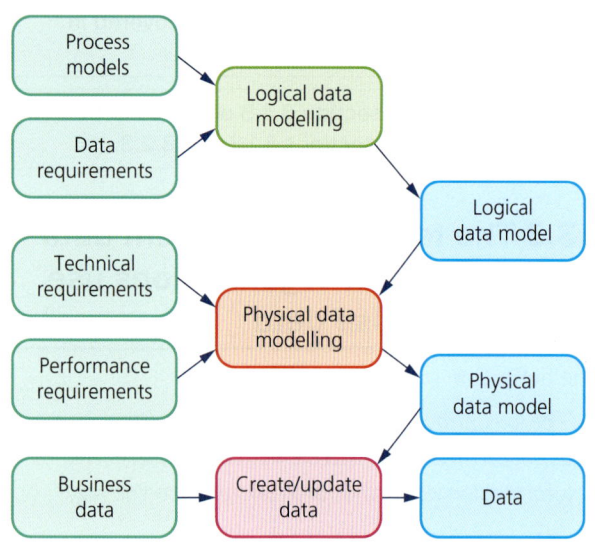

▲ Figure 6.8 Types of data models

ERD

An ERD can be used during creation of the conceptual, logical and physical data models. There are three main parts to an ERD including:

▶ entities – which will become the tables

▶ columns – which describe the tables and are also called attributes which will become fields in a database

▶ relationships – which are how the entities (tables) are linked. Relationships can be one-to-one (1:1), one-to-many (1:M, M:1), or many-to-many (M:M). These show the cardinality.

A completed ERD should not include any many-to-many relationships. If a M:M relationship does exist, then a link entity needs to be included.

For example, a shop sells dog food to its customers. The shop has a number of suppliers, each of which supplies many products. Many customers can buy many products. The ERD would look like this:

The M:M relationships need to be resolved. This is done by a link entity. In this case the link entity is a

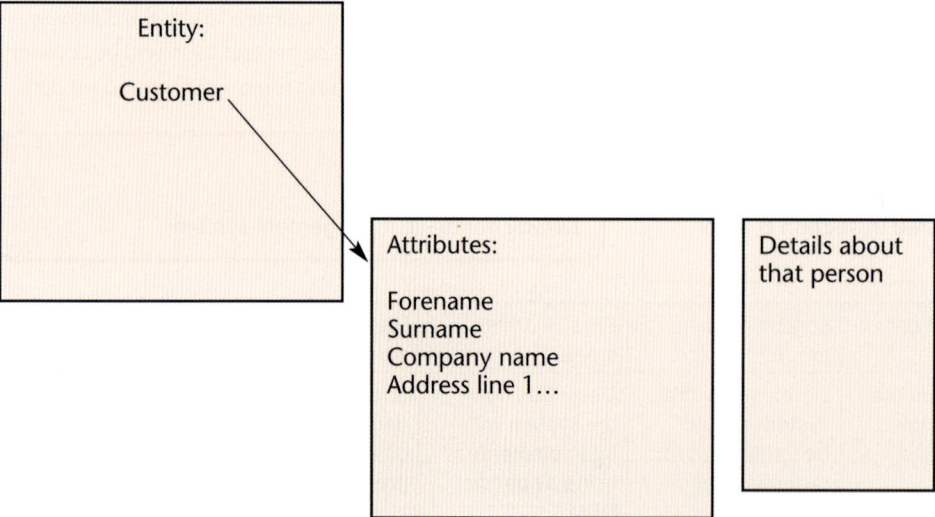

▲ Figure 6.9 First ERD for paint shop

CUSTOMER		ORDER		PAINT		SUPPLIER
CustomerID		OrderID		PaintID		SupplierID
Forename		CustomerID		Colour		Name
Surname		PaintID		Name		Contact Name
Address Line 1		Date		SupplierID		Phone
Town				Type		
County				Cost		
Postcode						
Phone						

▲ Figure 6.10 Resolving many-to-many (M:M) relationships for paint shop list of products from the product entity that a specific supplier can provide. It can be seen that the link entity contains the many relationships, but these are broken down into a 1:M relationship.

This stage of the ERD modelling is part of the conceptual data modelling process.

When the ERD is undergoing physical data modelling, it will be used to create the database. Primary and foreign keys will be used so that the links in the ERD can be created and to remove any data duplication. The ERD can be also known as relational database modelling.

As part of the logical data modelling stage, more detail is added to the ERD to show how the DBMS is going to be implemented. The ERD is developed to show the entity (table) name, attributes, **primary** and **foreign keys** and relationships defined.

A data dictionary is usually created as part of physical data modelling and will include details about each entity (table) and its attributes. The contents of the data dictionary will vary dependent on the software package that is to be used to implement the output of the logical data modelling.

The basic level of information that would be included in a data dictionary is shown in Table 6.7.

> Metadata is covered in section 6.4.3 of this content area.

Key terms

Primary key: a field in a table that allows each record to be uniquely identified. For example, every person 16 years or older in the UK has a National Insurance number. This uniquely identifies a person.

Foreign key: this is used to link tables together. A foreign key is a field in one table that is linked to a primary key in a different table.

Data dictionary

Data dictionaries contain data about the data that will be included in the database. It contains **metadata**.

Activity

1 Use the conceptual model ERD in Figure 6.9 and create the logical model ERD. Discuss your logical model ERD with the rest of your group. What are the similarities and differences?
2 Create a data dictionary for the logical model ERD you have just created.

Test yourself

1 Identify two types of data modelling.
2 What is a foreign key?
3 Identify and describe two components of a data dictionary.
4 How is the start of an Activity diagram shown?
5 What does an interaction diagram show?

Data	Description
Table name	The name of the table. A unique name for each table in the database
Field name	Each field is identified
Field data type	The data type allocated to each field: text/string/date/Boolean, etc.
Field length	The number of characters allocated for the contents of the field
Field default value	If a field has a default value that automatically appears on the creation of a new record
Field validation	Any validation applied to the field
Table security	Who has access to write, update, edit, delete etc. values to and from the table
Keys	Primary and foreign keys are identified
Indexes	Any field which is indexed
Relationships	Relationships between tables identified: one-to-one, etc.

▲ Table 6.7 The basic level of information that would be included in a data dictionary

6.4.1 Factors that determine how data is gathered, entered and maintained

Data can be gathered from a range of sources. When the data has been gathered it needs to be entered into the digital system so it can be used, processed and analysed. Data also needs to be maintained so it retains its usefulness.

The six Vs

You have already learned that big data is just that – a data set that is big.

Because big data is so big, a six-factor system has been developed to assess the quality of the data held within the big data set. These factors are known as the six Vs:

1 volume
2 variety
3 velocity
4 variability
5 value
6 veracity.

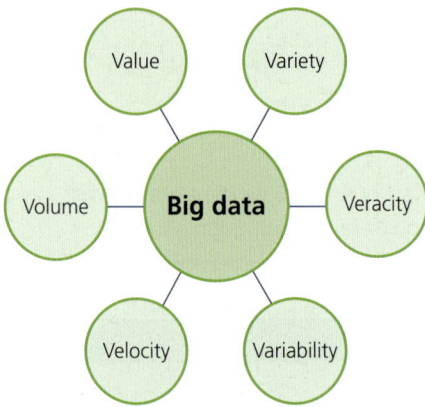

▲ Figure 6.11 The six Vs

> Big data, data warehouses and data lakes are covered in section 6.4.2 of this content area.

Volume

The volume factor considers the vast amount of data that is generated every second, minute, hour and day. The volume of big data is so big it is measured in terabytes (1,099,511,627,776 bytes) and petabytes (1,125,899,906,842,624 bytes). This volume of data can be stored in a data warehouse or data lake.

Variety

The variety factor considers where the data has come from and in what form. This is one of the most important factors of big data. Big data can be gathered from a range of sources and can be structured, unstructured or semi-structured. But, to process and analyse the data, it must be structured and organised. Big data can come in many forms, for example, text, images, voice, emails, videos, audio and geospatial data.

Velocity

The velocity factor considers the speed at which the data is being generated in real time. This can also refer to how quickly the data changes and moves from one point to another. Big data should be processed quickly as it is generated and gathered quickly. Big data never stops increasing in speed and amount so there is more data to process now than there was when you started to read this paragraph.

Volume, variety and velocity are the three main factors when considering big data. Big data includes very large amounts of data in a variety of structures which must be processed quickly. As the gathering and use of big data has increased, more V factors have been introduced.

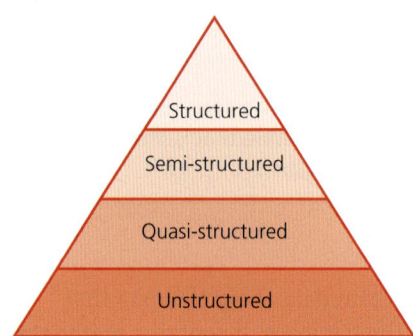

▲ Figure 6.12 Types of big data

Variability

The variability factor describes how quickly, and to what extent, the data being processed and analysed is changing. This is important because if data is constantly changing, then the results of the processing and analysing may not be the results when applied to the latest data. If the variability of data is high, then processing and analysing may have to be repeated as the results may not be up-to-date or complete.

Value

The value factor describes the meaning of the data after processing and analysis, and how much the results meet the predefined goal of the analysis.

Veracity

Another factor relates to how trustworthy the data is – its veracity. Veracity can have an impact on the level of confidence users can have when processing and analysing the data. The greater the volume of data collected the higher the probability that some of the data will be untrustworthy which, in turn, will decrease the level of confidence. As the volume, variety, velocity and value of data increases, the veracity factor will probably decrease.

> **Activity**
>
> Create an infographic aimed at 16 to 18-year-old students providing details of the six Vs of data.

Data assurance

Any data that is gathered, processed and analysed must have a high level of reliability and quality. Data that is unreliable and/or of low quality is bad data.

The reliability of data means how complete and accurate it is. For example, data about a customer without including their name would be unreliable and of low quality. This is because the record of this customer is incomplete and inaccurate.

To increase the quality of data, verification and validation techniques can be used.

> Validation and verification techniques are covered in section 6.3.1 of this content area.

Data redundancy

Data redundancy occurs when the same piece of data is stored two or more times on a digital system. This can be linked with data inconsistency. This is when the same piece of data is saved in different formats or data types across a digital system. Data redundancy can cause data inconsistency. This means that the results of processing and analysis can provide unreliable/ meaningless information. This in turn could lead to uninformed and incorrect decisions being made at strategic and tactical levels.

Another example of data redundancy is where data is stored in two different locations and is updated in one location but not the other. It is also possible that if data is stored in more than one location the data can be assumed to be unique instead of being duplicated. However, if comprehensive data modelling has been carried out, the risk of data redundancy occurring is reduced.

Data redundancy can also be useful to an organisation. A back-up copy of data can be classed as data redundancy but the back-up is vital to the smooth and continued functioning of an organisation.

> Data types are covered in section 6.2.1 of this content area.

Qualitative data and quantitative data

Data can be gathered using surveys and questionnaires. These may follow an interaction with an organisation as part of customer feedback. This enables an organisation to gather qualitative and quantitative data during the feedback. An organisation may also carry out research about a specific issue, such as feedback on a new concept.

> Qualitative data and quantitative data are covered in section 6.1.1 of this content area.

Research population

The organisation will select the research population it will use to gather the required data and information. The number of people required to complete the research is called the research population. How large this population is will depend on the defined requirements needed from the results of the research. When selecting the research population it is important to consider the demographics of the population, which must relate to the pre-defined requirements of the research. If the research population is skewed in terms of, for example, gender or age group, then the results of the research my not provide reliable or high quality data.

Legislation and regulatory compliance

It is important that the data being gathered by an organisation complies with up-to-date legislation and any relevant regulations. The legislation that should be complied with includes the DPA.

> Legislation and regulations are covered in Content area 4, section 4.1.

Ethics

Ethics has already been defined as the moral principles, or rules, that govern a person's attitudes and behaviour. One of the main factors to be considered is that of the owner of the gathered, entered and maintained data.

All data should be gathered, entered and maintained in the same way, with no bias or pre-conceived perceptions. It is important that the Equality Act and Data Protection Acts are also considered. Linked to these legislative Acts should be the principles that all data should be gathered, entered and maintained transparently, with accountability and fairness.

Organisations who gather, enter and maintain data should remember that the data owner is the person who owns the data, and they have rights under the DPA. One of these rights is that of privacy and security of their data. This also includes being informed, and to provide consent, if their data is passed on to a different business or organisation.

When data is being handled by an organisation, the data holder is entitled to view the algorithms that may be used when decisions are being made using their data. These algorithms must be created to ensure that no bias is evident. The data owners may also be permitted to have access to the data sets that were created using the algorithms. This would enable the data owners to notify an organisation of any errors and have them rectified. The UK government defines eight areas that should be considered. These relate to:

- personal data
- equality and discrimination
- sharing and reuse of data
- copyright and intellectual property
- freedom of information
- statistics
- information governance
- sector specific legislation.

Data Protection Act Chapter 4, section 1.2
Equality Act Chapter 4, section 1.4

Activity

In a group of three or four, investigate the UK Government areas to be considered when gathering, processing and analysing data. Use a range of websites to research further details for each of these areas. Create a digital communication and present your findings to the rest of your teaching group.

Organisational factors

There are organisational factors related to gathering, entering and maintaining data. These factors include time, skills and cost.

An organisation will need time to gather the data. This could be as a result of gathering data from external or internal sources. It is also possible that data included in a data warehouse or lake will need to be gathered, based on a specified need and requirement. There will be a financial impact in terms of how long it takes to gather the data.

When the data has been gathered it will need to be entered into a digital system. This again will cost an organisation in terms of time and skills and paying the employees to carry out this task.

To keep the gathered data useful, it must be maintained. This can mean carrying out regular maintenance to check that the data is still up to date. If the data relates to personal data, then it is a legal requirement that data is kept up to date. In addition, it is possible that data has become redundant, repeated or out of date over time. These tasks may increase costs to an organisation in terms of time, skills and paying employees to complete the task, including liaising with the data holders.

As the amount of data held by an organisation increases, so increased storage capacity may need to be sourced. This may take the form of servers either on-site or virtual in the cloud. This will, again increase costs to the organisation in terms of costs for storage, either a one-off purchase price or an on-going contractual payments and software. Staff will also need to be trained to handle the data. This may take the form of updating skills or updating knowledge relating to legislative requirements.

It is, however, possible that an organisation will contract an external, third party to carry out these tasks. This will have an increased financial cost to the organisation and may have other impacts in that if the needs and requirements are not clearly defined, the gathering of the data may not fully meet those needs.

Test yourself

1. How many bytes is a petabyte?
2. Variability is one of the six V factors. Identify three other six V factors.
3. What is the variability factor?
4. What is meant by the term 'data redundancy'?
5. What is meant by the term 'data reliability'?

6.4.2 The purpose of data analysis tools and their use in business

Data warehousing

Data is key to organisational decision-making. But this is often complicated by the fact that data is held in many places and in a variety of formats across the business.

The solution to this problem is a data warehouse: a central collection of key data, integrated into a pre-defined format. This allows the data to be used across the business to make business-critical and evidence-based decisions.

A data warehouse can be known by other names, as shown in Figure 6.13.

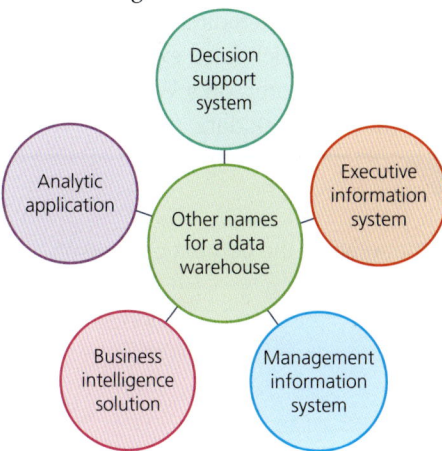

▲ Figure 6.13 A data warehouse can also be known by other names

Data in a data warehouse will be taken from a range of internal and external sources.

This data will therefore likely be in a range of formats and will therefore need to be subjected to **data cleaning**. This process converts the data into the standardised and pre-defined format of the data warehouse.

> **Key term**
>
> **Data cleaning:** the process of going through data looking for errors and correcting them, or excluding data where errors have been located.

It is important to understand that when data is stored in a data warehouse it does not change and cannot be edited. Data can be added to the data warehouse as the data sources are updated. Decisions need to be made about what data can be included and excluded from a data warehouse. Data is only input to a data warehouse when a use for the data has been identified by the organisation.

A data warehouse can be used to process and analyse data on previous data, rather than data that is being currently used and updated by an organisation. Data warehouses are used by specific business users to analyse and extract a particular meaning from the data that was defined when the data warehouse was set up. Analysing data held in a data warehouse usually focuses on changes in data in any given time period, for example analysing data on sales in the previous year.

Data stored in a warehouse must be secure, reliable, easy to retrieve and easy to manage to enable analysis to be carried out. The data warehouse can be stored on physical storage devices or in a secure area in the cloud.

Data lakes

Data lakes are data stores that hold data in an unstructured way. Unlike data warehouses there is no defined format to the way the data is structured; the data is stored in a raw state as it may not be processed.

Unlike a data warehouse where the sources and format of the data are defined, a data lake takes data from all sources in any data type. The data is only converted into a predefined format when it is ready to be used and analysed.

It is possible that some of the data stored in a data lake will never be used. This may be because the data lake contains all data collected over time in a raw and unprocessed state.

But one benefit of a data lake is that, because there is so much data held there, a range of analysis and processing can be carried out on the stored data.

Data stored in a data warehouse or data lake can be referred to as **big data**.

> **Key term**
>
> **Big data:** very large data sets that can be analysed to produce information such as trends and patterns. Big data cannot be analysed using traditional analysis data tools.

Data mining

Data mining is used by organisations to process and analyse raw data. Software is used to look for trends and patterns in big data sources such as a data warehouse or data lake.

Data reporting

Data reporting takes data and converts it into information. It can show an organisation *what* is happening now, rather than analysis which shows *why* something is happening. Data reporting includes the tasks of consolidating and organising the data, formatting and summarising the results. One output from data reporting can be to convert the data into charts and graphs to enable the end users to visualise and understand the results of the data reporting.

6.4.3 The role of metadata classification in defining the meanings of data

Metadata can be defined as 'data about data'.

Metadata can be classified in three main ways:
▶ administrative
▶ descriptive
▶ structural.

Administrative metadata provides administrative instructions about a file. The metadata provides details about any restrictions to be placed on the file. Administrative metadata could include access rights or a specific file type.

Descriptive metadata relates to how the data is identified. Descriptive metadata could include titles, dates or keywords. For example, the runtime of an audio file could be classified as descriptive metadata. Descriptive metadata is referred to as metatags when being used on a website.

Structural metadata describes how a digital asset is arranged or structured. The obvious example of structural metadata is a data dictionary.

Data dictionaries were covered in section 6.3.3 of this content area.

Many uses of digital systems have metadata attached to them. For example, emails, webpages and word-processed documents include metadata. A database will have a data dictionary which is an example of structural metadata. File Explorer, on a Windows digital system, also includes metadata.

6.4.4 The use of data/access entitlements/permissions management, and their impact on organisations and stakeholders

User access restrictions including usernames, passwords and passphrases, data access levels/permissions and multi-factor authentication are covered in Content area 8, section 8.2.3.

You have already learned that data is the most valuable asset of any business or organisation. It is important that data is protected against any unauthorised access or editing. Permissions can be set to limit what a user can do with data. Permissions are also known as **privileges**.

In addition, **rules** can be set. These rules can be part of the permissions and privileges to attempt to increase the data security. Examples of rules that can be set include access to all files, and access to specific files and folders. In an organisation, these rules are likely to be based on the job role.

If the use of data is not kept secure and access rules set, there is a possibility that data can be leaked or edited. The impacts of the loss or unauthorised editing of data can include:
▶ data bias
▶ inaccurate data leading to incorrect processing and analysis and uninformed decisions being made
▶ a data breach/leak leading to a possible legal and/or financial impact.

6.4.5 How data can be accessed and managed across different platforms

The role and use of APIs in managing, accessing and using data

While it is important that an organisation maintains the integrity and security of its data, it is also important that authorised users are able to access, manage and use it.

Users should be able to access data across a range of platforms and digital systems using a range of software applications. It is also important that users are able to connect to the data remotely and on-site.

> APIs are covered in Content area 8, sections 8.1.3 and 8.2.3.

An Application Programming Interface (API) enables users to manage, access and use data across a range of platforms. The main advantage of using an API is that users do not have to create accounts for multiple websites. Accessing a website through an API means that only an account for the API needs to be created. This means that users can access a range of websites, and data, using their log-in credentials for the API.

An API allows applications to access data and interact with other applications or systems by sending and receiving requests. To send and receive the requests, the API uses JSON.

> JSON is covered in section 6.2.1 of this content area.

An API sends a user's action (request) to a digital system. The digital system sends a response back to the user.

For example, a user adds a product to their shopping basket. The API will tell the website that a product has been added to the basket and the website puts the product into the user's virtual basket. The basket is then updated.

Figure 6.14 shows how this interaction occurs.

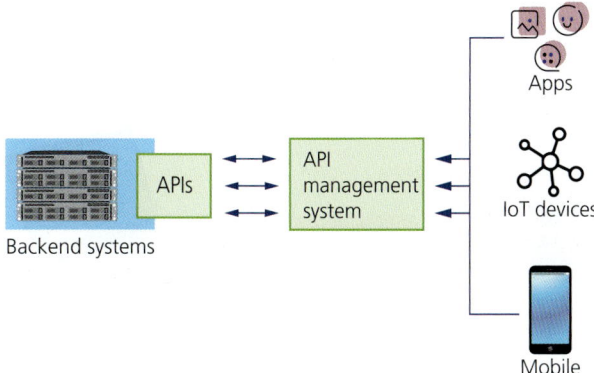

Backend systems • APIs • API management system • Apps • IoT devices • Mobile

▲ Figure 6.14 APIs

The API will ensure that authorised users can access and manage data but, to maintain the integrity of the data, the API must be maintained to ensure the highest level of security. It is also important that the certification of the API is appropriate. The certification should be set to the level which is appropriate to the data being accessed. This means that the API certification should be set to Partner or Private certification and not Public.

Another advantage of using an API to enable authorised users to access data is that the business can make changes to the backend systems, usually a DBMS. This will have no impact on the authorised users as long as the basic structure and behaviour of the API does not change.

APIs are most commonly used by the financial services sector, CRM systems and online retailers.

6.4.6 The concepts of data at rest, data in use and data in motion, and when each is used

Data is collected for use in many different situations. Data also needs to be stored, used and shared.

Data can be categorised into:
▶ data at rest
▶ data in use
▶ data in motion.

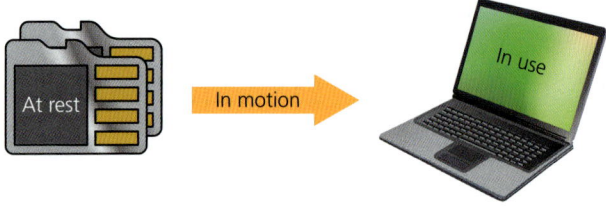

▲ Figure 6.15 The three states of data

Data at rest is data that is stored and not moving. Data at rest means the data is not moving from, for example, device to device or network to network, such as from an internal network to a cloud storage area.

The data can be stored on, for example, a hard drive, a cloud storage area, an external storage device such as a flash drive or any other storage device. Data at rest also includes data that has been archived.

Data in use is data that is being actively used and/or processed by a user or digital system. This can also be data that is being used by a digital device to function.

Data in motion is also known as data in transit. This is when data is moving from one location to another, and includes moving data via the internet or through an internal network. Data in motion can also apply to data that is being moved from a local storage device to a cloud storage device. Data in motion is considered to be less secure and more susceptible to security attacks.

Research

Research the data that is used by an online shopping website. Categorise the data as:
▶ at rest
▶ in use
▶ in motion.

Discuss your findings with the rest of your group.

The potential technical threats and vulnerabilities to data is covered in Content area 8, section 8.1.3.

The processes and procedures to mitigate threats and ensure security is covered in Content area 8, section 8.2.3.

Test yourself

1 Where can data at rest be stored?
2 Identify one example of data in use.
3 What is another term for data in motion?
4 How can data in transit be moved?
5 Which category of data is at most risk of a security attack?

Skills practice

A retailer with a physical presence is considering moving to an online presence. The retailer sells sports clothing and equipment. They are hoping that by moving to an online presence they will increase their customer base to worldwide. The retailer is considering carrying out analysis of pre-existing data sets to evaluate the feasibility of the move.

At the moment stock records are kept on a spreadsheet on a standalone digital device. The stock records are manually updated twice a week, with stock orders being completed once a week.

The retailer has been informed that to have a successful online presence, stock levels need to be updated automatically to ensure that customers are advised if an item they want to buy is out of stock. Customers should also be advised of when items will be back in stock. The retailer has also been recommended to require customers to register to use the online presence, providing their name, contact details and any default settings such as a 'leave in a safe place'.

You have been asked to:
▶ Describe the advantages of using big data sets to research the potential market for the online presence.
▶ Create a logical modelling ERD and data dictionary for the stock levels and reordering process.
▶ Create an online form for the customers to use when registering, showing the data types applied to each field and validation routines.
▶ Explain to the retailer how targeted marketing can be carried out using customer registration details.
▶ Provide details of how and why the retailer should implement a CRM system.
▶ Explain how customers could use an API to access the retailer's online presence.
▶ Explain the concepts of data at rest, data in motion and data in transit including any security implications related to these.

Assessment practice

1 Explain how data becomes knowledge.
2 Explain the differences between strategic and tactical decisions.
3 A driverless car uses sensors. Explain how these sensors can help the car to learn.
4 There are rules associated with a fixed-width data file. Identify and explain three of these rules.
5 Explain two advantages of presenting data in an infographic.
6 Identify and explain one impact of data redundancy.
7 Explain the advantages of creating a data model.
8 Explain, using an example, what is meant by descriptive metadata.
9 Identify and describe the three main six V factors when considering big data.
10 Describe the difference between a data warehouse and a data lake.

Content area 7: Digital environments

A digital environment is one where a wide range of digital devices communicate and support the content and activities within it. There are numerous components of a digital environment including the internet, the cloud, virtual environments, networks and physical environments. An organisation can implement a combination of these components to support its business functions. The components used depend on the size and function of the organisation, and whether it is based in one location or several locations, which may be globally.

In this content area you will learn about the various environments, what they are, how they work and their importance to a digital environment.

Learning outcomes

In this content area you will learn about:

- 7.1 Physical environments
- 7.2 Networks
- 7.3 Virtual environments
- 7.4 Cloud environments
- 7.5 Resilience of environment.

7.1 Physical environments

7.1.1 The features and characteristics of different types of physical computer systems

Personal computers

A personal computer is intended for personal use or by a small group of individuals. It is often shortened to PC. A personal computer consists of a CPU containing the arithmetic, logic and control circuitry on a single integrated circuit (IC). It contains RAM and read only memory (ROM). It contains a hard disk drive (HDD) and various I/O devices including display screen (monitor), keyboard and mouse. There are several characteristics that have made computers powerful and universally useful.

When selecting a personal computer it is important to consider what it is going to be used for, for example graphic design, running a home business, gaming, etc. Considerations include:

- **Size:** when considering the physical size of the computer, the amount of space available for it is clearly important. An 'all-in-one', that is a laptop where all components are located in the same casing as the monitor, may be appropriate if space is limited. However, these can also come in a variety of sizes depending on the size of the monitor itself. (For gaming, a larger screen is preferable.) Separate base units can come in a variety of sizes too, for example mini and micro, and require a separate monitor (or connection to a compatible television). Decisions therefore have to be made about where the base unit will be situated, for example on a desk, under a desk, etc.
- **Storage:** the optimum amount of hard disk storage capacity is dependent on what the computer is being used for, as well as how frequently it is used. The more data required to be stored, the more storage capacity required. While some home computer users store their data on the cloud, most people store their data on the hard drive. The speed of the hard drive is dependent on whether it is a solid-state drive (SSD) or a hard disk drive (HDD). SSDs are not mechanical and are faster than HDDs. SSDs are measured in MBps (megabits per second) which is based on the speed of the data passing through the microchips, while HDD are measures in rpm (revolutions per minute), which is the number of revolutions of the disk in one minute.

- **Processor speed:** the clock speed of the processor determines the speed in which the CPU (Central Processing Unit) retrieves and interprets instructions to carry out the tasks. Faster processor speeds increase the efficiency and productivity of the computer system.
- **RAM (Random Access Memory):** this is the temporary storage for the system and is known as volatile. Once the computer is shut down, any storage within the RAM is lost. The size and type of RAM has an impact on the response speed of the computer. Gamers would look for RAM size in excess of 16 GB, whereas people who work a lot with editing media files will need a RAM size of 32 GB or more.
- **Internet connectivity:** some computers provide the option to connect to a wired network or a wireless network. The importance of this is dependent on the type of network connections that the user has available to them.

Mobile devices

A mobile device is a general term for a small, handheld computer. This includes, but is not limited to, tablets, e-readers, smartwatches and smartphones. Mobile devices often have:
- a rechargeable battery that powers the mobile device
- cellular or Wi-Fi access to the internet
- the ability to connect to other devices, for example via Wi-Fi or Bluetooth
- a small size and light weight
- a display screen
- a touchscreen or keyboard that acts as the input device
- the ability to download data and information including apps, photographs, books, music and so on.

Servers

A computer server can be hardware or software and is used to provide centralised services such as data and/or resources such as printers and applications, to other devices (known as clients) on a network system. There are different types of server, for example database, mail, print, file, etc.

The client connected to the network sends a message to the server, for example sending a request to access email (from a mail server) or accessing an application (from an application server), and the server then responds to the request from the client, providing them with the data and/or resource requested.

Research the given types of servers and complete the table:

Type of server	Purpose of server	How it works	Features	Characteristics
File				
Application				
Database				
Print				
Virtual				
Mail				
Web				
Domain Name Server (DNS)				
Hybrid				

Smart/internet enabled devices

These are devices that are part of the IoT. They are nonstandard computing devices that are connected wirelessly to a network and can transmit data. Examples of such devices are described here.

▶ **Connected appliances** – for example refrigerators, lights, smart speakers, washing machines.
▶ **Smart security systems** – wireless cameras and doorbells which record movement and allow remote monitoring.
▶ **Wearable health monitors** – can be personal monitors such as the Fitbit, or monitors used by the health sector to monitor the heart rates, insulin levels and so on of patients.
▶ **Wireless inventory trackers** – using the IoT an organisation is now able to monitor its inventory in real time. Using wireless inventory trackers provides real-time communication and location

monitoring, which is precise, detailed and itemised (this can include stock coming in, stock on-site, deliveries to customers), and touchless data collection using scanners allowing items to be tracked and recorded. It assists warehouse management by encouraging the efficient use of space and the activities that take place in the space by considering the identified usage pattern. In addition, the integration with robotics saves time and money.

After completing the activity, have a group discussion about your research. Consider whether other members of the group have listed advantages and disadvantages that you haven't considered.

For each of the smart/internet enabled devices, research how they are used by different industry sectors, and the advantages and disadvantages of using them for the sector. List the results of your research in a table.

Smart/internet enabled device	Industry sector	How it is used	Advantages	Disadvantages

7.1.2 The features and characteristics of hardware and peripherals used in physical computer environments

Input devices

An input device is any hardware that sends data to a computer. It enables the user to interact with it and control it. The user may be another computer or a measuring device, as well as a human.

Input device	Types and features
Keyboard	**Types:** • USB – does not require additional power source, not affected by external sources/signals, cyber criminals would need to install a keylogger to access the information being typed in. • Wireless – can be placed in any convenient location on the desk, reduces workplace clutter. • Integrated (e.g. integrated into laptops) – smaller than a standard keyboard, more mobile than external keyboards. • Onscreen (e.g. smartphones, tablets, touchscreen devices) – more mobile than external keyboards. **Features:** • Users input commands and controls the computer. • Used to input letters, numbers and symbols. • Includes additional keys, e.g. **function keys**, **control keys**, **special purpose keys**.
Mouse	**Types:** • USB – does not require additional power source. • Wireless – can be placed in any convenient location. **Features:** • Allows user to control the onscreen cursor. • Contains two buttons (left and right button). • Tracker wheel to scroll up and down pages on websites and documents.
Scanner (converts a document or image/ photograph into a digital file)	**Types:** • Flatbed scanners – the scanning mechanism rolls under the document/image, etc. to capture the image. • Sheetfed scanners – only useful for single sheets. The sheet is fed through the scanner and moves along the scanning beam. • Integrated scanners – used in **automated teller machines (ATMs)** for cheque processing and approval. • Drum scanner – used for capturing a picture or transparency at a very high-resolution rate. They are very expensive and considered a major upgrade by an organisation. Used a lot in photography and the film industry. • Portable scanners – designed to capture text and other data. Depending on where they are used, the data can be stored on the device and transferred to a computer at a later date or can transfer the data remotely automatically. Portable scanners are powered by batteries which are re-chargeable. Portable scanners can also include mobile devices such as mobile phones.
Digital cameras	**Types** • DSLR (Digital Single Lens Reflex) – a mirror is used to reflect the light so that the image can be seen in the viewfinder (as with a film SLR) camera but the image is stored digitally, e.g. on an SD card. • Compact – a fixed-lens camera that is small in size (sometimes referred to as a pocket camera) – stores the images digitally on a media card such as an SD card. **Features:** • Used to capture images and videos independently of the computer system. • Photos and videos can be transferred to a computer system using a cable (usually USB cable) or by removing the memory card and inserting into a card reader linked to the computer. Can also transfer using Bluetooth capabilities.

Input device	Types and features
Microphone	**Types**: • Inbuilt into the monitor or a computer system. • Part of a headset containing earphones and microphone. **Features**: • Captures audio sounds which are transferred to a computer system and converted to a digital format. • Can be used to record or relay audio sounds as part of web conferencing, or audio and video streaming.
Graphic tablet (sometimes referred to as a digitiser)	**Types:** • Screenless graphics tablets – these tablets do not have a built in display and must be plugged into a computer system in order to view the image. • Hybrid graphics tablets – paper is placed directly onto the tablet and images can be drawn without the need for a computer screen as the user can see the image they are drawing on the paper. **Features**: • Converts hand drawn artwork and/or text to digital images. • Users draw on a special flat surface using a stylus (which is part of the graphic tablet system). • The image appears on the computer screen and can be saved, edited and/or printed. • More versatile than a scanner as the user can see the 'live' image appear on the screen as it is being created.
Touchscreen	Commonly used for smartphones, tablets and laptops, although there are now many desktop monitors which have touchscreen facilities. **Features**: • Allows the user to point, drag or select options on the screen without using a mouse. • Touch sensitive so reacts to fingers moving across the screen. • Contains two, sometimes three, simultaneous touchpoints (depending on the number of people using the touchscreen at the same time or the requirements of a particular package). Uses of multiple touchpoints include to zoom and use two finger taps.
Sensors	Sensors are a form of input device to a computer system. A simple classification of sensors is that they are either **active** or **passive sensors**. A sensor can also be classified by the detection methodology that it uses, for example electrical or chemical. They can also be classified per the way that they work with data – **analogue sensors** and **digital sensors**. **Types:** • temperature sensor • humidity sensor • pressure sensor • seismometer • breathalyzer • smoke detector • touch sensor • vehicle speed sensor.

▲ Table 7.1 Input devices and their types and features

Key terms

Function keys: arranged in a row at the top of the keyboard, they are assigned a unique meaning and used for a specific purpose.

Control keys: provide cursor and screen control. Include four directional arrows, Home, End, Insert, Delete, Page up, Page down, Control (Ctrl), Alternate (Alt) and Escape (Esc).

Special purpose keys: include Enter, Shift, Caps Lock, Num Lock, Spacebar, Tab and Print Screen (PrtScn).

Automated teller machine (ATM): An automated Teller Machine is a special computer that allows a bank account holder to manage their account.

Active sensors: sensors requiring an external signal or a power signal.

Passive sensors: do not require external power signals and directly generate an output response.

Analogue sensors: produce a continuous output signal relating to the quantity being measured.

Digital sensors: work with discrete digital data. The digital data is used for conversion and transmission.

Output devices

An output device is any type of peripheral that receives data and either displays it, projects it or produces a hard copy of the data reproduction.

Printer

▶ **Inkjet printer** – sprays streams of quick-drying ink on paper. The ink cartridges are usually disposable and separate for each of the main colours (cyan, magenta, yellow and black).

▶ **Multifunction printer (MFP)** – sometimes referred to as an All in One (AIO) printer that includes a printer, a fax and a scanner. Because it combines a number of devices it takes up less space and requires fewer cables because it is only one device.

▶ **3D printer** – used by industries such as aerospace, archaeology, dentistry, information systems and biotechnology. A physical object is created from a digital model using a layering technique. They can use plastics, polymers, metal alloys and even food.

▶ **Plotter** – used for printing vector graphics using, for example, a pencil, pen or marker to draw multiple continuous lines. They can work on very large sheets of papers or on any flat surface, for example plywood, sheet steel and aluminium. The same pattern can be drawn thousands of times without any degradation to the image. They are much larger than traditional printers and usually more expensive.

Monitor

Modern monitors are created using flat-panel display technology which are backlit with light-emitting diodes (LEDs). It interprets and displays graphical output signals from the computer's graphics card.

There are two main types of monitors, LCD (Liquid Crystal Display) and LED (Light Emitting Diode). While both types of monitor use liquid crystals to help create the image, the difference is the backlights that are used. LED monitors use LEDs as the backlights while LCD monitors use fluorescent lights. LED screens also use less power than LCD monitors and are therefore more environmentally friendly. Not all LED monitors provide the same image quality as it depends on the layout of the LEDs. If the LEDS are evenly placed across the entire screen they provide a better quality than when the LEDs are only placed around the edge of the screen. So, depending on the LED configuration, for example the edge configuration of the LEDs, there are instances when they do not provide such a good quality image as an LCD monitor. With LCD monitors the liquid crystals are placed between two sheets of glass for every pixel. The current from the monitor liquifies the crystals allowing the white light that is emitted from the backlights to pass through them.

LED monitors that are configured with LEDs across the entire screen will have a better viewing angle spectrum than the edge configuration. This is because the visibility of the edge configuration reduces as you move away from the central position of the screen. There are also more glare issues with edge configuration than with the full screen LED configuration. This is where an LCD monitor will win over an edge configured LED monitor as the LCD will have better viewing angles and less glare because all areas of the screen are lit.

The edge-configured LED monitor is the thinnest and cheapest option but, as previously stated, it has its disadvantages. If space is at a premium, they are worth considering. As with other components, it depends on what the monitor is being used for primarily. For example, gamers should purchase an LED monitor with the LEDs placed across the entire screen because of the various viewing angles available.

Interactive whiteboard

These are like touchscreens in that they are both an input and an output device. Some are like a large touchscreen TV and others are operated using a touchpad which is connected to a computer. The computer is then connected to a projector. Some interactive whiteboards are independent touchscreen devices that perform different operations and carry out commands. The end user can alter the different information present in the board which can even be copied and transferred to the next slide in a presentation. Data/information that is presented on the board can be selected, moved, copied, deleted and analysed. They can be written on like any other whiteboard using special pens. The software with the whiteboard detects the pen movement to turn this into text.

Projector

A projector takes images from a computer or DVD/CD player and projects them onto a screen, wall, or any other flat, lightly coloured surface. They can display slides or moving images and are often about the size of a toaster weighing just a few pounds. In recent years, handheld projectors have been developed.

Headphones

Sometimes referred to as earphones, headphones enable the user to listen to audio outputs and are often used during online calls and conferences. Headphones can also have built-in microphones, in this case the headphone is an output device and the microphone is an input device. Many headphones are wired, which means that they connect to each other, as well as the computer or mobile device, via a wire.

There are many styles of headphones and earphones available. Over-ear headphones are composed of two speakers that sit over a user's ears, with soft padding to ensure a comfortable fit. These two speakers are connected by a band over the user's head.

Earbuds, however, fit inside the user's ears and are wireless. These come in lots of shapes and sizes, so a user can choose the fit which suits them best. Earbuds are often adjustable thanks to additional soft, rubber ear inserts that are provided by the manufacturer.

Recent advances in technology have led to a variety of innovations in headphones, such as better sound quality. There are also now options of noise-cancelling headphones – noise from the outside world is blocked out, to immerse the user fully in the sound – or bone-conducting headphones – these sit to the side of the ear, meaning that the user is still able to hear sounds from the outside world, allowing awareness of surroundings, for example while out running.

Speakers

A speaker is an output device that converts the electrical signal produced by a computer or mobile device's soundcard into sound. Speakers can have a wired connection, where leads are connected to the output line of the digital device, or have wireless connectivity. Many devices such as laptops, computers, smartphones and SatNavs have inbuilt speakers, although external speakers can be connected for improved sound quality. Speakers are rated via the **frequency response**, **total harmonic distortion (THD)** and **watts**.

> ### Key terms
>
> **Frequency response:** the measurement rate of the highs and lows of the sounds produced by a speaker.
>
> **Total harmonic distortion (THD):** the amount of distortion created by the signal amplification.
>
> **Watts:** the total amount of amplification available for the speakers.

Sound card

A sound card enables sound to be heard through speakers or headphones. There are two types of sound card, half-duplex which is purely an output device and for listening to sound, and full duplex which also allows sounds to be recorded via a microphone (therefore it is an input and output device).

A sound card is also referred to as an audio card or audio output device and can be either in the form of an expansion card or built into the motherboard (the OSC, onboard sound card). There are also sound cards available that connect via USB and are therefore external to the computer. The type of sound card selected will depend on what it will be used for. Some sound cards will have onboard processing power and therefore will not impact on the CPU or RAM of the computer.

Graphics card

A graphics card processes images or videos. It is sometimes referred to as a video card and can be onboard (in the same way as the sound card) or included as an expansion card. The type of graphics card will depend on what it will be used for. For example, graphic and web developers as well as gamers usually select a dedicated graphics card in the form of the expansion card. They have their own memory and processing power to meet the processing requirements for graphics and videos. Many of the higher specification graphic cards have **VRAM (video random access memory)** which is important for gamers and graphic designers as it helps to improve the quality and reduce the loading times of the image.

> ### Key term
>
> **Video random access memory (VRAM):** a dual-ported memory. It is an expensive form of RAM which has the capability to perform reads and writes simultaneously. VRAM can also be accessed by two devices at the same time and is commonly used to increase the speed of video cards.

Processors

The processor is commonly called the central processing unit (CPU) and is often identified as the brain of the computer. It is the software that is the brain, and the CPU is an extremely efficient calculator. A CPU is very good at working with numbers, but if it was not for the software, the CPU would not know how to do anything else. The CPU contains the Arithmetic Logic Unit (ALU) that carries out the arithmetic, logic and decision operations (executes commands). The control unit (CU) directs the operations of the processor (decodes instructions into commands).

Characteristics

▶ **Clock speed** – PUs contain a clock (like a clock that you can hear ticking hanging on a wall or sitting on a shelf). Every time the clock ticks, an instruction is carried out. Depending on that the instructions are that are being carried out, they may only require one tick of the clock, whereas more complex and/or larger instructions may require numerous ticks. This means that the faster the clock speed, the more instructions that can be carried out per second.

▶ **Cache** – The cache is similar to RAM in that the CPU gets instructions from it. Cache is faster than RAM when it comes to reading from it and writing to it, and therefore temporary data that is frequently used is stored in it. This means that applications are able to load faster and even work 'offline', without having to be downloaded every time they are needed. As a result of these efficiencies, a higher cache value helps to increase the speed of the computer.

▶ **Cores** – A core is a processor with its own cache. Modern computers have a CPU containing multiple cores, for example quad-core. This enables each core within the CPU to carry out a different process which in turn increases the speed of the overall system and the computer is able to 'multi-task'.

▶ **Bit size** – A bit is one binary digit and the smallest unit of data. It can either have a binary value of 1 or 0. The instructions and data accessed by the processor is in binary code. In terms of the processor, the bit size (number of bits) relates to the registry size and the size of the data that it can work with at any one time. A 64-bit processor is not necessarily faster than a 32-bit processor as there are other things that increase the speed of a computer, for example its clock speed.

Some of the different types of processors are presented in Table 7.2.

Processor type	Uses	Characteristics	Features
Microprocessor	• Computers • Smartphones • Vehicle speed controllers • Traffic lights • Military applications	• Consists of a CPU • Uses an **external bus** to interface with RAM, ROM and other peripherals • Expensive and complicated, with many instructions to process	• Provides inbuilt monitor/debugger program with interrupt capability • Offers parallel I/O (Input/Output) • Instruction cycle times • External memory interface

Microcontroller	• Mobile phones
	• Vehicles
	• CD/DVD players
	• Washing machines
	• Security alarms
	• Lighting systems
	• Fire detection systems
	• Keyboard controllers
	• Watches
	• Cameras
	• Microwave ovens
		• Consists of three things integrated on one chip:
		  – CPU
		  – Memory
		  – I/O
		• Uses an internal controlling bus
		• Inexpensive and straightforward, with fewer instructions to process than other types of processors
			• Contains a processor reset
			• Program and variable memory (RAM) I/O pins
			• Device clocking central processor
			• Instruction cycle timers

Processor type	Uses	Characteristics	Features

Embedded processor (can be confused with microcontrollers)	Controls electrical and mechanical functions	At a basic level, a CPU chip placed in a system that it helps to control
		• Similar functions to microcontrollers but it:
		  – integrates with the system it is part of in a different way (embedded processors require additional resources such as RAM and registers in order to control a system, whereas a microcontroller contains everything it needs in one single chip to carry out the same function)
		  – can also perform different functions (an embedded system is used to create an automated device as well as controlling devices using a microcontroller)
		• Requires external components to function, e.g. peripheral interfaces
		• Often a component within a microcontroller
			• Simple design
			• Limited computational power
			• Limited I/O capabilities
			• Minimal power requirements

Digital signal processor (DSP)	Used for measuring, filtering and/or compressing digital/analogue signals such as voice, audio, video, temperature, pressure or position and manipulates them mathematically Examples include: • speech processing • image processing • medical processing • biometric processing • seismology • radar	• Processes signals such as voice, video, temperature, audio, position and pressure from the real-world in real time that has been digitised. It then mathematically manipulates them so that they can be converted to another form of signal or display information. E.g. digital TVs use DSP to ensure that the TV is compatible with different types of video standards, e.g. VGA, SVGA, SXGA, UXGA • Have programmable processors so that their parameters can be changed to accommodate different applications, e.g. for use in cell phones, digital TVs or sound cards • Can quickly perform mathematical calculations such as add, subtract, multiply and divide	• Analyses and manipulates signals • Can process signals using a computer, **Application Specific Integrated Circuit (ASIC)** or **Field Programmable Gate Array (FPGA)**

▲ Table 7.2 Processor types and their uses, characteristics and features

Key terms

External bus: also known as an external bus interface (EBI) or expansion bus. It is a type of data bus that enables external devices and components to connect with a computer.

Application Specific Integrated Circuit (ASIC): a microchip designed for special applications such as a handheld computer.

Field Programmable Gate Array (FPGA): an integrated hardware circuit that can be programmed to carry out one or more logical operations. Uses include the rear viewing cameras on cars, data analytics, encryption, compression and AI (e.g. within deep neural networks).

Memory

Memory type	Characteristics
Random Access Memory (RAM) – main memory	• Fast and long life • No need to refresh • High power consumption • Expensive • Data is stored electrically on transistors • Volatile (if a computer loses its power, all data in the RAM is lost) • Stores the operating system, applications, and graphical user interface (GUI) • Volatile memory can be changed, upgraded or expanded easily by users • Faster than secondary memory
Read Only Memory (ROM) – main memory	• Cannot be changed by a program or a user • Does not depend on an electric current to save data • Data written to individual cells using binary code • Is non-volatile (retains data in memory after computer is switched off) • Used for parts of the computer that do not change, e.g. initial boot-up sequence of the software • Faster than secondary memory
Cache (memory in the CPU)	• Speeds up access to data and instructions stored in RAM • Holds frequently requested data and instructions so that they are available for the CPU when required • Two types: internal cache (within the processor), external (on the motherboard) • Faster than main memory • Shorter access time than main memory • Stores data for temporary use

Secondary storage devices	• Different storage media which is attached to a digital system, e.g.
	  – hard disk drives HDD
	  – solid state drives (SSD)
	  – optical drives (CD/DVD)
	  – flash drives (USB/pen drives)
	  – memory cards (commonly used in cameras and handheld devices)
	• Non-volatile
	• Large storage capacity
	• Cheaper than primary memory
	• Can be portable or internal to the digital system
	• Slower than primary memory

▲ Table 7.3 Memory types and their characteristics

Motherboard/mainboard

Motherboards have different form factors (size, shape) and are the main printed circuit board within a computer and contain the buses (these are the electrical pathways). Components such as the CPU, heatsink and fan assembly, RAM, BIOS, chipsets, sockets, expansion slots, internal/external connectors and ports are all located on the motherboard.

Cooling

Cooling is required to keep several components cool within a computer system:
▶ CPU – CPUs generate heat and therefore need to have a cooling system in place to maintain their performance. CPUs can be cooled through air and liquid. For air cooling, thermal paste is applied to the heat spreader of the CPU, a heat sink is situated directly on top of the thermal paste and heat is extracted from the CPU through the thermal paste and into the fins of the heat sink. A CPU fan is attached to the heat sink and draws air through the fins. For liquid cooling, a layer of thermal paste is placed between the CPU and the base plate of the water block. The heat of the CPU is absorbed by the water which then travels through one or more tubes to what basically looks like a radiator. Fans are attached to the radiator and helps to draw the heat away from the cooler. The water then

syphons back to the water block and continues the process again
▶ graphics processing unit (GPU)
▶ RAM, which usually has its own metal heatsink to disperse the heat
▶ power supplies, which contain fans (**active cooling**)
▶ motherboards, which have heatsinks for components that get hot (and some motherboards also contain heatshields).

Test yourself

1 Compare and contrast two different types of scanners.
2 Identify three characteristics of a processor.
3 Describe the term 'form factor'.
4 Identify two peripherals that can be classed as an input and output device.
5 Discuss the different types of processors and their uses.

7.1.3 The purpose and functions of software used in computer systems

Operating systems (OS)

Type	Purpose	Functions/Features
Batch OS	Collects programs and data together in a batch before processing.	• Processes batches of data at regular intervals. • No user interaction required. • Batch processing carried out when least demand for processing power, e.g. at weekends or at night. • Can be set to run at specific times e.g. at the end of the month for a payroll system. • Batches processed on a first-come-first-served basis.
Multi-tasking/ time-sharing OS	Enables the use of a single computing resource for multiple uses at the same time.	• Involves the processor carrying out multiple tasks at a time, e.g. a user using the internet and also using a word-processing package. • The programs being used are either waiting, runnable or running. • The OS schedules the processes to be executed by the CPU (when to change between the processes based on their waiting, runnable, running status). • As the OS facilitates multi-tasking, several applications can be stored in the RAM at the same time complex to set up.
Real-time operating system (RTOS)	Switches between tasks rapidly giving the impression that programs are being executed at the same time. It aids the management of different hardware resources and hosts the applications that run.	• Typically used for embedded applications, i.e. systems within another application, e.g. a car management system. • Data is processed as soon as it is received by the processor. • It is high performance – fast response time and based on user requirements. • Priority scheduling – will process high-priority assigned tasks first. • Higher security and reliability standards – used for critical systems, e.g. aircraft controllers. • It will always produce the same output if the same input is used. This is known as determinism.

Type	Purpose	Functions/Features
Network operating system (NOS)	One of the most important types of operating systems, it runs on a server and provides the capability to manage users, groups, security applications, data and other networking functions. It facilitates share file and printer access across multiple users.	• The centralised servers are stable. • The security of the network is managed by the server. • Facilitates remote access to the servers from different locations and types of systems. • Upgrades using new technologies and hardware are easily integrated into the system. • The purchase costs and running costs are high. • There is a dependency on a central location. • Regular updates and maintenance are required.
Mobile OS	This is the software platform to run mobile devices. It is responsible for determining the functions and features available on the mobile device, e.g. email, text messaging, synchronisation with other apps, keyboards, etc. The mobile OS also determines which third-party applications (known as mobile apps) can be used on the mobile device.	Types of mobile OSs include: • Android OS – Google's open and free software stack that includes an OS, **middleware** and key applications for use on mobile devices (including smartphones). • iPhone OS/Apple IOS – only available on devices manufactured by Apple. It is derived from Apple's Mac OS. • Windows Mobile (for Windows phones) – is a Microsoft OS used in smartphones and mobile devices with or without touchscreens. Based on the Windows CE 5.2 kernel.

▲ Table 7.4 Operating systems: their purpose and functions/features

Utility software

Utility software is used to help manage, control and maintain the computer system and associated resources. Various operating systems have in-built utility software, for example disk scan or disk defragmentation, but there are also many programs available that can be installed onto a system. These include anti-virus and anti-malware software, registry cleaners and so on.

Application software

These are specific programs that are installed onto a computer system or made available via the cloud, for end-users to perform a range of tasks. Application software can include graphics and CAD software, office suite software such as Microsoft 365, and browser software such as Google Chrome and Firefox.

Code development tools (Integrated Development Environments, debuggers)

These are tools used by programmers when developing software. It enables them to create, maintain and debug the software they are developing.

Code development tools are often referred to as software development tools. Some of the tools include:
▶ **Integrated development environments (IDEs):** – this is a software application that enables software

developers to have access to the main developer tools within one central interface. IDEs will include as a minimum a source code editor, compiler and debugger. Depending on the type of IDE installed, they can include other tools such as class and object browsers and class hierarchy diagrams.
▶ **Source code editor** – a form of text editor which makes it easier for the writing and editing of code.
▶ **Compiler** – used to convert source code into a format (usually machine code) that a computer can execute.
▶ **Debugger** – all software coding can contain errors and bugs, and a debugger is used to test the code and identify potential errors and bugs. It may highlight code that contains an error or lines of code that could not be executed.
▶ **Assembler** – a program that converts assembly language (which is a low-level programming code) into machine code.
▶ **Linker** – sometimes referred to as a link editor. It is a program that collects and maintains object files and combines them into one executable file (program).

Test yourself

1. Explain the term 'RTOS'.
2. Describe how a batch OS functions.
3. Identify three code development tools.
4. Explain the purpose of utility software.
5. Discuss the role of a network operating system within a business. Consider the functions it performs and how this can benefit a business.

7.1.4 The benefits and drawbacks of software, hardware and peripherals in different contexts

In the previous sections you read about the features and characteristics of different computer systems, peripherals, hardware and software. Not all components which make up a digital system are suitable for all businesses, their environments or the industry sector in which they sit. Therefore, it is always important to consider the benefits and drawbacks of each identified component to ensure that it is appropriate and will produce the benefits that the business requires. In order to do this, you must understand the terms benefit and drawback.

Sometimes people can confuse a feature with a benefit, and they are different.
- A **feature** is something that a digital system, hardware/software component or peripheral has associated with its functionality.
- A **benefit** is the outcome/result from using the digital system, component or peripheral.

A **drawback** is something which can have a negative effect on or disadvantage a business. It may be that a specific piece of software may greatly benefit a

business, but the drawback is the cost needed to upgrade its current digital systems in order to use it.

Activity

An estate agent has offices in several different towns. Many of the valuers, who visit the properties that people want to sell, work from home. They only have to attend the office once a week. They use digital technology to upload the information and photographs of the properties that they have valued to the relevant office. The administration staff prepare the property details and photographs for display on the boards in the window and inside the office, and prepare advertisements for local newspapers and for social media. In addition, they arrange the viewings for the properties for sale.

Prepare a report on the types of computer systems, hardware, software and peripherals that would be required by the business. Explain the benefits and drawbacks of the components you have selected.

7.1.5 How physical data storage and recovery systems work, their features, benefits and drawbacks

Redundant array of independent disks (RAID)

This is the combining of multiple disk drives into a single unit (array). Because the disks work in unison, RAIDs are more reliable and faster, reducing data loss and improving overall performance. The increase in speed and reliability depend on the type of RAID used.

Types of RAID

- RAID 1 – this is sometimes referred to as disk mirroring. RAID 1 consists of two identical sets of disks (known as drives). Data is stored identically on both sets of disks and is used for **redundancy**. If one of the drives fails, then a business is still able to operate as it will use the other drive that is still functioning. The entire drive or disk within the drive can easily be replaced and the entire data from the still-functioning drive is copied across. There is an increase in the **read performance** due to the data being able to be accessed and read from either of the two functioning drives. There is however, a slightly higher **rate latency** as the data must be written to both drives in the array.

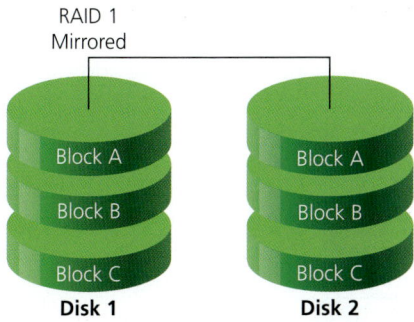

▲ Figure 7.1 Example of the configuration of RAID 1 storage

▶ RAID 5 – this contains at least three drives (although it can contain more) with blocks of data being striped across the multiple drives. Parity bits (correction codes) are also written across each drive after each sequence of saved data. A **parity** checksum is stored on one drive and used to calculate the value of the parity bits to check whether any data is missing and replace it. RAID 5 usually has its own hardware controller. Due to the **data striping**, it has an even more increased performance because the data is spread across the drives, and they can all be read at the same time. RAID 5, on the other hand can only withstand one disk failure and therefore other means of data backup should be implemented.

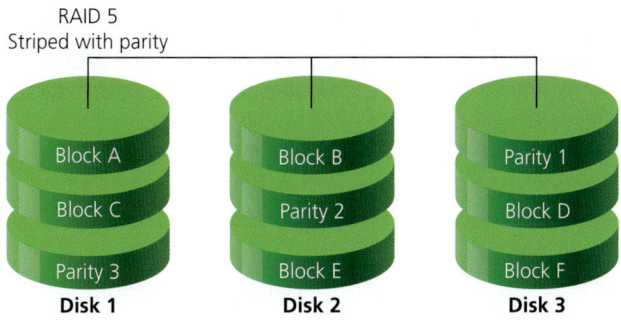

▲ Figure 7.2 Example of the configuration of RAID 5 storage

▶ RAID 10 – this is a combination of disk mirroring and disk striping and requires at least four drives and a disk controller. The four disks are divided into two sets of two in a RAID 10 configuration. RAID 10 has the benefit of data redundancy due to there being a mirror image of the data and increased read and write performance because of data striping. It does not, however have parity checking. There is the disadvantage however, that should there be a total failure in one of the sub-arrays, the entire system can crash. The advantage is that the rebuild time is reduced in the event of a disk failure as long as there is a mirror image still available.

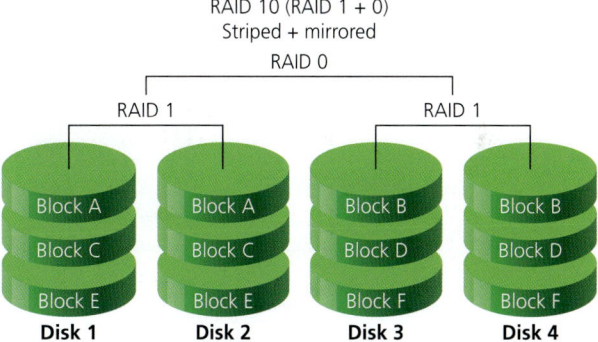

▲ Figure 7.3 Example of the configuration of RAID 10 storage

Network attached storage (NAS)

This is a storage device that is connected to a network and acts as a central point for the storage, management and access of files. As the NAS is connected directly to the network, it can only be accessed by authorised networked devices and users. NAS is configured with data transfer protocols (DRPs) for example, NFS (Network File System) allowing the transmission of data between devices.

One of the main benefits of a NAS is the fact that additional storage can be added to it easily. To increase the storage capacity, there is just the requirement to add additional disk drives. Data recovery and backup is easier with a NAS, but it should never be the only backup option used as accidental deletion, failure and/or virus infection cannot be overcome because there is no inbuilt option to recover deleted files. Security is always a major concern where data is involved and, like any other system/device that is used, it is always important that they are regularly maintained and updated and that any system connected to the same network, are carefully monitored and controlled.

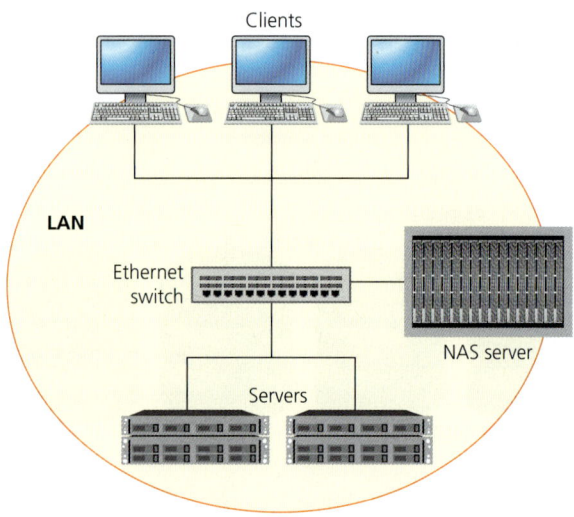

▲ Figure 7.4 Example of a NAS configuration

Storage area network (SAN)

A SAN is a network of interconnected storage devices accessible by computers and servers. Its purpose is to store, manage and protect data. SANs use block storage where data is broken down into 'blocks' and are stored separately. Each block is allocated a unique identifier and a software program reassembles the requested blocks. Once a request is made, the software identifies the relevant blocks based on their unique identifier and reassembles them into one file that is then accessible for the user. The block storage system means that data can be accessed more quickly than a standard file-storage system and can be access using different types of operating systems.

A SAN network is usually connected using fibre optic cabling which is faster than other forms of cabling and uses a protocol known as a fibre channel, providing better performance. SANs are expensive in relation to the technology required and it is also complex to set up, configure and maintain. Invariably, this means

that there is the additional cost for a skilled network manager to monitor and maintain the SAN. SANs, however, are more easily scalable by adding additional hard disks and switches.

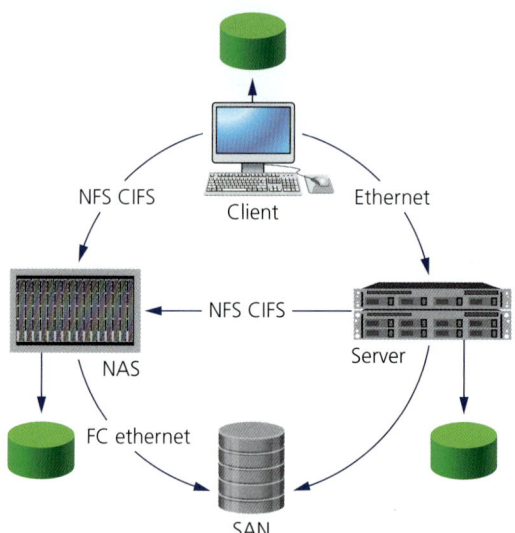

▲ Figure 7.5 Example of a SAN configuration

Test yourself

1 Compare and contrast RAID 1, RAID 5 and RAID 10.
2 A government research department uses vast quantities of data (big data). What type of physical storage should they consider implementing and why?
3 Explain the role of NAS.
4 Identify one benefit and one drawback of RAID 1.
5 Describe two benefits of NAS to an organisation.

7.2 Networks

7.2.1: The benefits and drawbacks of connecting devices to form networks

Networks are used to connect electronic devices so that they can communicate and share resources. Connecting devices to a network can have benefits as well as drawbacks. Some of the benefits and drawbacks are also dependent on whether the network is wired or wireless, as well as the network model being implemented.

Benefits	Drawbacks
Reliability – if the network has a central server, the data is stored on the server. If one of the devices connected to the network fails, the data is still available via the central server.	**Servers** – server-based networks are reliant on servers such as file and mail servers. It is therefore important that networks have powerful computers used as servers which makes it easier to set up and maintain the network. If the main server fails, the entire network can fail, and the users are unable to carry out their tasks.
Accessibility on a global scale – if the network is a WAN or has access to a WAN through the internet, then resources such as data, files, information are accessible to the user of a connected device irrespective of where they are situated in the world. This results in business processes not suffering delays waiting for files, data, information, messages to 'arrive' at their destination.	**Implementation costs** – although networks can reduce costs, the initial implementation of a network can be costly, e.g. the cost of additional hardware such as servers, switches, hubs and routers.
Manageability – for server-based networks, individual software applications do not need to be installed on every computer connected to the network. The software is installed and can be upgraded/updated centrally and then be readily accessible to all connected devices.	**Maintainability** – networks need to be maintained and this tends to involve complex configurations and installations. These tasks can only be completed by experienced network specialists who can demand high wages for their skills and knowledge.
Security – security is controlled centrally to mitigate risks against security breaches.	**Security** – networks can present many security risks, especially for large networks where there are many users accessing and sharing resources and files. A virus, for example, can easily be spread across all devices connected to the network. Security risks is covered in detail along with how they can be mitigated in Content area 8, page 216.

▲ Table 7.5 Benefits and drawbacks of connecting devices to form networks

7.2.2 The features, characteristics, benefits and drawbacks of wireless connection methods

Wireless connection means that devices can be connected to the internet or a network without using any form of cabling. They contain access points that are connected to a router, hub or switch using an Ethernet cable. These access points distribute the Wi-Fi signal to the devices on the network. The access points and the wireless devices must be configured with the same **SSID (Service Set Identifier)**.

> ### Key term
>
> **SSID (Service Set Identifier):** a unique name identifying the name of the wireless network.

Characteristics of wireless connection methods

A good wireless network has:
- **Scalability** – the network needs to be able to increase in size without major re-configuration of the entire network and with minimal disruption.
- **Load balancing** – the network must be able to handle high volumes of demands and therefore

sufficient access points should be incorporated so that if one access point is overloaded, then the network will automatically focus the traffic in the direction of another access point.
- **Performance measurement** – the ability to measure the performance of the network from the end-user's perspective. This requires the real-time monitoring of users and the devices and applications that are in use. It can also monitor in real time the status of the various network components and the potential impact on any devices being used.
- **Application and web content filtering** – all networks must mitigate against threats and therefore application filtering is used to protect organisations/users from potential malicious threats and to prevent performance issues.
- **Prioritisation of applications** – there is a need in all business environments to guarantee the performance levels of applications that are critical to the functioning of the business. The applications that are high priority are therefore prioritised.
- **Network Access Control (NAC)** – it is important that there is a mechanism in place that can register and secure any device that is not owned by the organisation controlling the network. The NAC controls the role of the user and enforces any policies associated with accessibility, etc.

▶ **Network management systems** – this enables network managers to monitor, maintain and optimise the network. They are usually a single application specifically designed for network management.

▶ **Mobile device management** – this is a software application used by the network managers and their teams to secure, monitor and manage the mobile devices that have access to the network and use multiple service providers and operating systems.

▶ **Role Based Access Control (RBAC)** – this is a system to restrict system access by assigning specific roles to users and the devices. When a role has been assigned, permissions and privileges are set for that particular user and/or device.

Benefits and drawbacks of wireless connection methods

Benefits	Drawbacks
Allows users to have access to data and information in real time while moving from location to location within the access area of the network without losing connection.	Can present security issues, for example unauthorised access to the network signal.
Provides scalability functionality as it is easy to configure additional devices and software.	A BYOD (bring your own device) policy, if an organisation has one, prevents central control over whether a device is infected, which can result in the inadvertent transfer of viruses and malware when connected to the network.
Installation of a wireless network is quicker and easier because there is no requirement for the installation of cabling (apart from the cabling required by the main components of the network such as routers and access points). The client devices on the network do not need additional cabling.	A persistent hacker can crack the code of a security protocol such as Wi-Fi Protected Access (WPA), allowing them to access the network and sensitive information.
Updating a wireless network is simpler and easier than a wired network. Devices can be updated via software updates.	There is the potential for interference of other wireless signals from other wireless networks used in the same proximity, e.g. another organisation in the same building.
If there are areas of the building where the installation of network cabling would be prohibited or difficult to address, a wireless connection enables network connection for these areas.	Managing and controlling the network can be difficult if people can connect their own devices and the network is left open.

▲ Table 7.7 Benefits and drawbacks of wireless connection methods

Test yourself

1 Identify two things that a wireless connectivity method should guarantee.
2 Explain the security issues when using wireless connectivity methods.
3 Describe RBAC.
4 Identify three features that a wireless connectivity method should have.
5 Identify and explain one benefit of using a wireless connectivity method.

7.2.3 The features, characteristics, benefits and drawbacks of wired connection methods

Characteristics of wired connection methods

Wired connection methods use cables to connect devices to the network and/or internet. There are different forms of cabling, including:

▶ **Cable** – this uses a type of coaxial copper cable and technology called **Data Over Cable Service Interface Specification (DOCSIS)**. Cat 5/Cat 5e (Category 5) cables are commonly referred to as Ethernet cables and used for providing the wired connection between devices and a network. Cat 5e has a higher throughput speed than Cat 5 and is also backward compatible.

▶ **Digital Subscriber Line (DSL)** – uses standard phones lines to send and receive information. This enables the user to make telephone calls as well as access the internet and transfer data.

▶ **Fibre optic** – this is the fastest method of delivering electrical signals by converting them into optical signals, transmitting them through a thin glass fibre and reconverting them to electrical signals. Fibre optic is a very reliable and secure transmission media that supports high bandwidths and can cover long distances.

Benefits and drawbacks of wired connection methods

> **Key term**
>
> **Data Over Cable Service Interface Specification (DOCSIS)** – a international telecommunications standard permitting broadband data transfer using the same cable systems that were used for transmitting cable television signals.

Benefits	Drawbacks
Security can be stronger for a wired network because the network can be configured with firewalls and other similar security applications that prevents unauthorised access.	Wired networks are more time consuming to install than wireless networks. This is because of the requirement to install the cabling and the connection of routers, switches and hubs, all of which will require configuring as well.
Unauthorised users cannot connect to the network without the use of an Ethernet cable.	There is less mobility with a wired network, e.g. if a person wants to take their device to the other end of the building, they will need to ensure that there is an access point for them to connect their device to.
Wired networks can be more reliable and stable than wireless networks. This is because the routers, hubs, switches, etc. are connected using physical cables and therefore the entire system is more robust. In addition, there is less chance of interference from other network signals that are in close proximity.	All networks require maintenance, but a large network will have one or more servers. The more devices that are connected, the more there is a need for additional requirements such as a server to manage the additional capacity on the network.
Wired networks are usually faster than wireless networks as each cable connected to a device transmits at the same speed.	Users wanting to access the network must have physical access via a cable as opposed to with a wireless network where they only have to be in the proximity of the network.
Wireless networks can suffer from 'black spots' where signals are inaccessible or difficult to get through (e.g. very thick walls in a building). Wired networks do not suffer the same problem as there is always a connection via a cable.	

▲ Table 7.7 Benefits and drawbacks of wireless connection methods

7.2.4 Different types of network

There are basically three types of networks: Local Area Network (LAN), Wide Area Network (WAN) and Personal Area Network (PAN). These networks have a wired connectivity option as well as a wireless connectivity option:

▶ **LAN** – a LAN is one of the most common and simplest types of network. It is the connection of groups of computers and low voltage devices across a short distance. A short distance can be within the same building or between a group of buildings that are in close proximity to each other and are used to share information and resources. LANs use routers to connect to WANs so that information can be shared, and data can be transferred quickly and securely.

▶ **Wireless Local Area Network (WLAN)** – these are basically the same as LANs but without the cables connecting the hosts and the servers. They are connected using wireless technology, commonly referred to as WiFi. Radio signals are used as the medium of communication with computer systems requiring wireless network cards. Most systems connect to the network using a router to communicate with other devices/systems on the network or for accessing the internet which is a Wide Area Network (WAN). The range of a WLAN can be within a room, a building or across buildings in close proximity to each other.

▶ **Wide Area Networks (WAN)** – cover a wide area for communication between computers. The internet is a good example of a WAN. A WAN comprises a series of LANs that have been joined together. A router is used to connect the LANs to WANs. Many WANs like the internet are not owned by one person or organisation but accessed by many individuals and organisations. Multinational organisations, however, invariably have their own WANs.

▶ **Wireless Wide Area Network (WWAN)** – use mobile phone signals as opposed to radio signals. The mobile phone signals are provided by mobile phone service providers and facilitate the transmission of signal over a wide area.

- **Personal Area Networks (PAN)** – very small networks of connected devices within one building and used by an individual or small business. PANs can be created using Ethernet cables, USB and/or Wirefire. They have limited distance accessibility; hence they are small and not used by larger organisations. Think of a PAN as a network within a single room for a single user (or users if they are in the same room), whereas a LAN is a network for a building for multiple users.
- **Wireless Personal Area Network (WPAN)** – the wireless version of PANs; they use Bluetooth technology or WiFi and have a limited distance of accessibility.

Test yourself

1. Identify the three main types of network.
3. Explain the term 'WAN'.
2. Describe the difference between a LAN and a WAN.
4. Identify the type of network that uses mobile phone signals provided by specific mobile phone service providers.
5. Compare and contrast the different types of networks.

7.2.5 Bandwidth and latency, and their effect on the performance of networks and connected systems

Bandwidth

Bandwidth is the measurement of how much data can be transferred from one point to another on a network over a specific period of time. When referring to the bandwidth of internet connections, it relates to how much data can be downloaded to a device from a server on the internet. The larger the available bandwidth, the greater the amount of data that can be sent and received across the network.

Bandwidth is not the same as the speed of a network. Speed is the rate at which data can be sent across a network. The bandwidth is the capacity (volume) available for that speed.

Bandwidths have limits and some devices use greater bandwidths than others. It is therefore always important that the bandwidth of the network can handle the bandwidth demands for all devices so that the overall speed of the network is acceptable for all intended users.

Bandwidths can be symmetrical or asymmetrical. Symmetrical bandwidths means that the same amount of data can be transmitted in both directions (referred to as uploads and downloads). Asymmetrical bandwidths means that the upload capacity is invariably smaller than the download capacity.

It is important when installing a network to calculate the amount of bandwidth required by identifying the maximum number of network users and multiplying the number by the bandwidth require for each of the applications that will be used. To calculate the bandwidth required for the cloud, it is important to know how much bandwidth is required to receive and send network traffic from the public clouds. The capacity can be affected by congestion on the connections used to reach the public cloud providers, particularly if the data is transmitting over the internet.

These steps must be carried out to calculate the bandwidth requirements for a specific application:
- Find out how much network bandwidth is available. This is expressed in bytes per second (Bps).
- Find out the average usage required by each application, also expressed as Bps.

Latency

Latency is the measure of the time taken to transmit data from one point to another over a network. Think of when you are accessing the internet and you are searching for information, the latency (delay time) is how long your browser takes to display the results of your search. Latency is measured in milliseconds (ms).

There are several factors that have an impact on the latency of a network:
- **Distance** – this is the distance from the device that is requesting the data, for example a computer to the server where the data is held. If the user accessing data from a server situated in another country, then the response time will take longer than if they were requesting the same data from a server based in the next town (or even the same country).
- **Transmission medium** – DSL, cable and fibre optic have a lower latency than satellite. Although the radio signals being sent to and from satellites travel at the speed of light, they still have a longer distance to travel than DSL, cable or fibre optic and therefore, due to this distance, the latency is much higher.
- **Routers** – a router takes time to process the information contained within the header of a data packet. Therefore latency (delays) occurs. It is therefore useful to install multiple routers, so allowing for router hopping to alleviate the delays.

If one router is busy, then the data packet hops across to the next free router.

- ▶ **Propagation delay** – this is the length of time it takes the data packets to travel from the device, for example laptop/computer, to the server.
- ▶ **Storage** – if the storage location is taking time to process the input/output requests because it is dealing with multiple requests, then a **bottleneck** is created and the latency can be high.

> Data packets are covered in Content area 8, section 8.2.3.

To summarise, an efficient network will have a high bandwidth and low latency. Bandwidth can be increased but latency must be considered as it can limit the amount of data throughput due to bottlenecks. So, bottleneck issues must be considered and rectified to maximise the potential of the bandwidth.

Key term

Bottleneck: limitation of data flow by network resources.

Test yourself

1 Bandwidth is the speed of the network – true or false?
2 Describe the term 'latency'.
3 Identify what makes a network efficient allowing for maximum throughput.
4 Explain how the size of the network's bandwidth has an effect on the sending and receiving of data.
5 Identify four potential causes of high latency on a network.

7.2.6 Different network models

Client–server

The client–server model is made up of a server with the computers on the network connected to it. These computers are referred to as the clients. The client computers and server communicate with each other through the local network or via the internet. The server is used to provide services and resources to the client computers. Client–server networks are usually used for web services, games hosting and private networks in organisations.

In order to access the services/resources each client has to have authorised access, that is a username and password, levels of accessibility, etc. In addition, the clients on the system are centrally controlled which means that the control of all connected devices is easier. Because of the architecture of this type of network model, it also means that if the server fails in some way, all clients on the network also suffer from the disruption. It is also important to remember that if there are a lot of clients on this type of network it can result in network traffic congestion which can slow the system down.

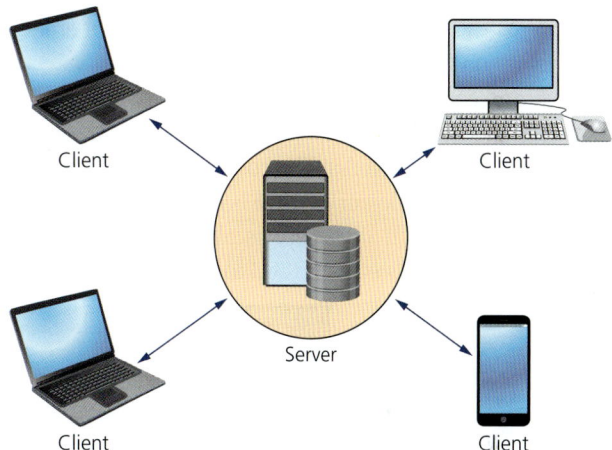

▲ Figure 7.6 Example configuration of a client–server network model

Thin client

Unlike a standard computer, a thin client runs by accessing resources on a central server. They do not have local storage such as hard drives and have a 'thinned' version of the operating system. Application servers are used to perform most of the data processing. As the resources are accessed from a central server, the costs are lower. It is easier to manage these network models with respect to carrying out upgrades and setting security requirements as this all takes place on the server as opposed to the individual thin clients. Thin clients invariably do not have any means to save data anywhere other than the server nor do they allow the user to install unauthorised software. It is simple to increase the size of the system (scalability) by adding additional thin clients and setting them up to be recognised by the central server.

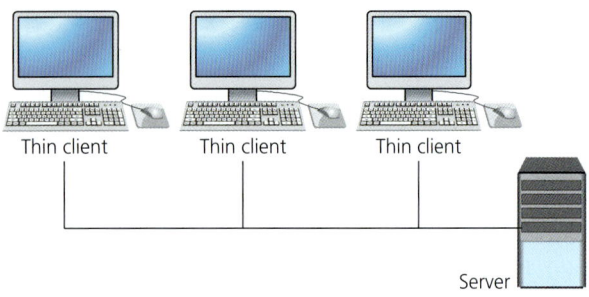

▲ Figure 7.7 Example configuration of a thin client network model

Peer-to-peer (P2P)

This type of network model does not have a central server. Each computer system is a 'peer' to the other computers on the network. The 'peer' acts as a client and a server, allowing every user to share folders, files and peripherals. Each device on the network provides resources to the network and consumes resources that the network provides. This results in a more improved performance than client–server networks. The cost of a peer-to-peer network is obviously lower as there is no additional cost for a separate server. It allows easy scalability because an additional computer is easier to add to the network.

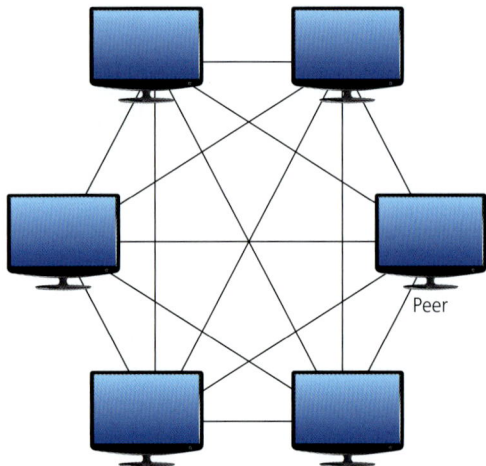

▲ Figure 7.8 Example configuration of a P2P network model

Test yourself

1 Compare and contrast the three different types of network.
2 Explain why a P2P network provides a higher performance than a client–server network.
3 Describe one benefit of a thin client network.
4 Identify two benefits of a P2P network besides higher performance.
5 Why is a client–server network more secure than a P2P network?

7.2.7 The characteristics of network topologies

Logical versus physical

Physical topology is the layout of devices on network, that is the physical structure of the network including devices, cables, switches, hubs, routers and so on. The logical topology is the way the data passes from one device to another through the network or how it acts on the network media regardless of the physical interconnection of the devices. Logical topologies are linked to the protocols that direct how data moves across a network. Protocol is the agreed format for transmitting data between two devices.

Star

This is where each device (or node), for example a workstation, server or printer, is connected to a central device (known as a hub, switch or concentrator) using a point-to-point link.

The central hub receives signals from a node and transmits (repeats) the signal to all other connected nodes on the network. The hub manages and controls all functions of the network.

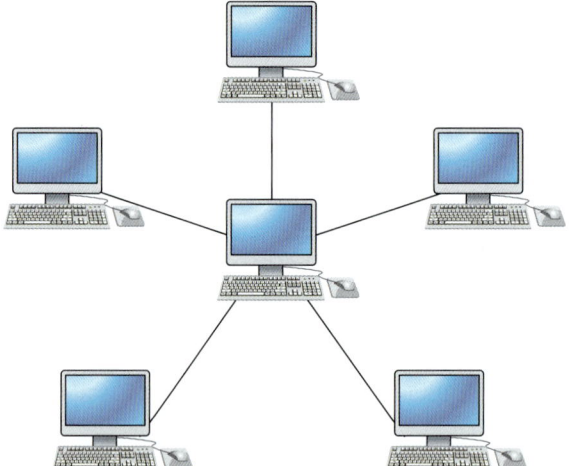

▲ Figure 7.9 A star topology

Advantages of star topologies

▶ Easy to install, wire and reconfigure.
▶ No disruption to the network when connecting or removing nodes.
▶ Easy to troubleshoot. This is because all data goes through a central point known as an 'intelligent hub'. These can be used to monitor and manage the network.
▶ Faulty equipment such as network cards, media and nodes can be easily isolated.

Disadvantages of star topologies

▶ If the hub fails, the network fails.
▶ It requires more cabling than many other topologies.
▶ It can be expensive (mainly due to the cost of the hub).

Mesh

A mesh topology is commonly used in WANs and is often found in public networks, such as the internet. A mesh network requires every device to have a point-to-point contact with every other device on the network. This is known as a fully connected mesh topology.

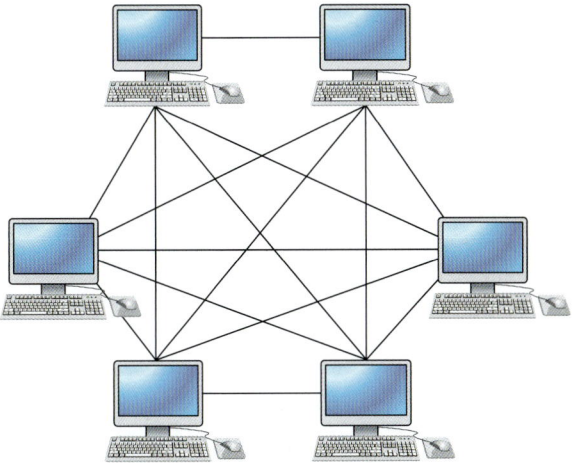

▲ Figure 7.10 A mesh topology

This fully connected approach is impractical, and a 'hybrid' approach is therefore often used, wherein only the most critical devices, such as critical mainframes, are interconnected. This is sometimes referred to as partially connected mesh topology.

Advantages of mesh topologies

▶ No traffic problems occur as there are dedicated links.
▶ It is more robust than other topologies. If there is a failure in one of the links, it does not impact the whole system.
▶ Data travels down a dedicated line and is therefore more secure.
▶ Fault detection is easier as there are point-to-point links.

Disadvantages of mesh topologies

▶ There is, quite literally, a 'mesh' of wiring, which is difficult to maintain.
▶ As there is a connection between every node, installation is complex and costly.

Tree

A tree topology (also known as a star **bus topology**) is a combination of star and bus topologies. Groups of star-configured nodes are connected to a linear bus **backbone cable**.

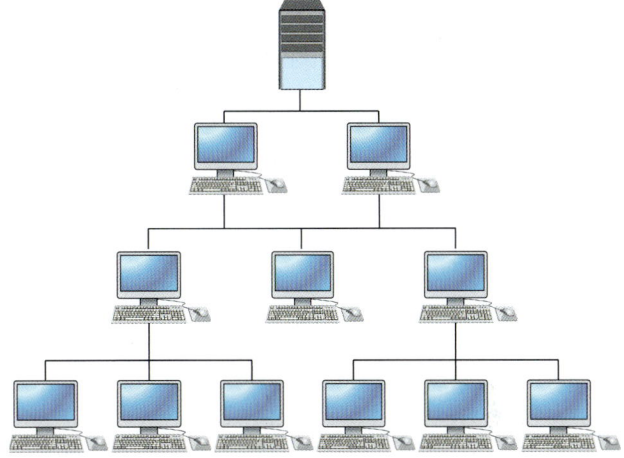

▲ Figure 7.11 A tree topology

> ### Key terms
>
> **Backbone cable:** the part of a network that combines different networks into a single complete network. The backbone carries the bulk of the network traffic.
>
> **Bus topology:** all nodes are connected directly to a central cable that runs up and down the network – this cable is known as the backbone. Data is sent up and down the backbone until it reaches the correct node.

Advantages of tree topologies

▶ There is point-to-point wiring for individual segments.
▶ It supports the expansion of existing networks.
▶ The network can be reconfigured to meet existing needs.
▶ Other nodes on the network are not affected if one node fails.
▶ It provides a hierarchical and central data arrangement of the nodes.

Disadvantages of tree topologies

▶ The length of the network depends on the cable used.
▶ The network is dependent on the backbone cable – if that fails, then the whole network fails.
▶ The bigger the network, the more complicated it becomes to configure.

- There is a limit to the length of each segment within the network that is also dependant on the type of cabling used.
- Performance can be slow due to the number of nodes on the network.
- Network traffic can be heavy because data travels from the centre cable.
- Costly and time consuming to implement with complex maintenance requirements.

Virtual Local Area Network

A VLAN (Virtual Local Area Network) is the method used to logically separate out networks. Imagine a LAN for a business with multiple departments or even multiple geographical areas. A VLAN is used to separate each department's area on the network into a virtual network. It helps to increase the efficiency of what would be a very large LAN and saves on network resources. It also helps to reduce the time taken for the transmission of data packets (latency).

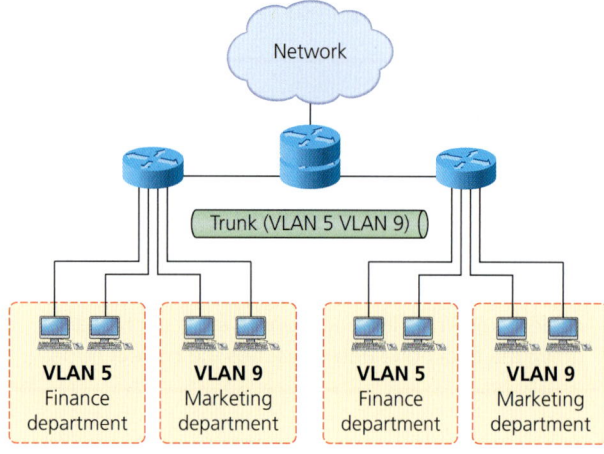

▲ Figure 7.12 A VLAN

Advantages of VLANs

- It provides more security control as each VLAN is a simulated/virtual separate network within a larger overall network. Therefore, sensitive information is not accessible by areas of the larger LAN/WAN.
- Latency is decreased (the data packets are transmitted around a smaller network area).
- It is easier to scale upwards or downwards as each area can be addressed in isolation and not impact other areas on the main network.
- It is easier to troubleshoot problems on a smaller network than a larger one.

Disadvantages

- It can be more expensive to implement as additional routers may be required to control the traffic of a very large network.
- Maintenance, including the addition of extra equipment, needs to be carried out using a logical and structured process to maintain existing VLAN segmentation.
- This means that implementation planning can take longer than other, simpler set-ups.

Test yourself

1 Describe the differences between a logical and a physical network.
2 Identify one advantage and one disadvantage of a mesh topology.
3 An organisation wants to create a network for various departments which are based in different geographical locations. Identify which network topology would be the most appropriate and justify your response.
4 Describe the configuration of a star topology.
5 Compare the two different types of mesh topology.

7.2.8 The role and characteristics of common components of a network

Server

The role and characteristics of different types of servers were discussed in section 7.1.1 of this content area.

Internet connection/internet backbone

Internet service providers (ISPs) are companies who privately own the core networks (known as the internet backbone). The ISPs have contracts with each other to connect their networks on a grand scale to form a large global network. Businesses and individual users will hold a contract with an individual ISP that will allow them to access the entire internet and interact with other businesses and individuals regardless of their location. ISPs use the transport layer and internet layer of the TCP/IP model to make computer-to-computer connections and transmit the data between them.

Further information on the TCP/IP model can be found in section 7.2.10.

There are three tiers of ISPs:
▶ Tier 3: these are the local ISPs such as the ones that you have a contract with for internet access.
▶ Tier 2: these are ISPs who act as a go-between Tier 1 and Tier 3 ISPs. They are regional or country-wide, for example Vodafone. They act in the same way as Tier 1 ISPs but have a wider reach with respect to internet access. Tier 2 ISPs will charge a fee to Tier 3 ISPs for accessing their systems.
▶ Tier 1: these are the ISPs that have the global reach across the internet. They charge the ISPs from the Tier 2 category for using their internet access.

Router

A router is used to manage communication between different devices on a network. It also allows any connected devices to access the internet at the same time.

There are wired and wireless routers. A wired router requires devices to be connected using Ethernet cables. A wireless router allows for wireless connectivity but will invariably have some wired ports also available. Routers operate within the network layer of the OSI model and use the physical, data-link and network layers. The router will use the IP address of the connected devices on the network to deliver data packets to the correct device. Routers are known as self-learning devices as they will automatically identify new devices connected to the network and will ensure that the additional connected devices also know of its existence. Routers support filtering and **encapsulation**.

> Further information on the OSI model can be found in section 7.2.9.
>
> The OSI model is discussed in section 7.2.9 of this content area.

Key terms

Encapsulation: is where additional information is added to the transmitted data as it travels through the different levels of the OSI or TCP/IP layers. When it arrives at its destination, the reverse process happens: de-encapsulation. The additional information is removed so that the receiver only receives the data.

Media access control (MAC): this is a code which is in the device's network interface card (NIC), identifying the physical device.

Network switch

A network switch is not required for internet access, whereas a router is. Network switches are used for large business networks and data centres where the devices are connected using Ethernet cables.

A network switch is used to connect devices within a network. Unlike the router, a network switch will only transmit data to the device that the data is intended for, while the router transmits to multiple connected devices including other networks. A Layer 2 switch uses the **media access control (MAC)** address of each connected device so that it can send the data to the correct device and works in the data-link area of the OSI model. A Layer 3 switch works in the network layer of the OSI model and uses the IP addresses of the destination device. Network switches that operate in the network layer of the OSI model are usually used to support VLANs.

The simplest way to think of a network switch is that it only routes the network traffic to a particular destination as opposed to the port of every connected device.

Client

> Clients were discussed in detail in section 7.2.6 of this content area.

Test yourself

1 Identify three different types of servers and explain their function.
2 Explain the function of a network switch.
3 Identify the layer of the OSI model that a router functions in.
4 Explain how an internet backbone functions.
5 Identify two characteristics of a router.

7.2.9 The seven-layer OSI model to describe how applications communicate over a network, including the function and related protocols of each layer

The OSI model was developed by the International Organization for Standardization (ISO) in 1977. It was designed to show how a network system communicates and operates.

> The ISO is covered in section 4.2.4.

The OSI model divides the complex task of networking into layers. This allows someone to work on the design and debugging of one layer without affecting the others.

There are seven layers with the OSI model stack. They are split into two groups:
▶ upper layers.
▶ lower layers.

7	Application	Upper layers
	Network process to application	
8	Presentation	
	Data representation and encryption	
9	Session	
	Interhost communication	

▲ Table 7.8 OSI model stack (upper layers)

4	Transport	Lower layers
	End-to-end connections and reliability	
3	Network	
	Logical addressing	
2	Data link	
	Physical addressing	
1	Physical	
	Media, signal and binary transmission	

▲ Table 7.9 OSI model stack (lower layers)

Each layer contains a different group of tasks required for a network to communicate. It must be remembered, however, that not all network systems implement layers using this structure.

The software in the upper layers performs application-specific functions, for example data formatting, encryption and connection management. Upper layer technologies include HTTP (Hypertext Transfer Protocol) and SSL (Secure Sockets Layer).

The lower layers provide functions such as routing, addressing and flow control. Lower-level technologies include TCP (Transfer Control Protocol), IP (Internet Protocol) and Ethernet.

The OSI model simplifies how network protocols are designed. It was designed to ensure that different equipment, such as adapters, hubs and routers, are compatible regardless of which manufacturer builds them.

Application layer

This is the interface between the end-user and the network and provides support services to applications requiring network services. The most utilised service is file transfer because different file systems often use entirely different naming conventions and syntax data. The application layer must overcome these issues. The types of applications that reside in this layer include Google Chrome and Mozilla Firefox.

The protocols associated with this layer include:
▶ FTP (File Transfer Protocol)
▶ WWW (World Wide Web)
▶ HTTP
▶ TCP
▶ NFS (Network File System).

Presentation layer

This layer translates data into suitable formats so that it can be read (or understood) by the application. The presentation layer supports data compression, provides security through data encryption and determines the structure of the data. It communicates through gateways and application interfaces and uses services such as FTP and NFS.

The protocols associated with this layer include:
▶ JPEG
▶ MIDI (Musical Instrument Digital Interface)
▶ MPEG
▶ many other music and picture formats.

Session layer

This layer allows applications running on different computers to communicate with each other. The connection is commonly known as the 'session'. The session-layer process is:
▶ establish the session
▶ manage data transfer
▶ tear down the session.

The session layer provides a synchronised service where checkpoints are inserted into the data stream. If there is a problem during transmission, only the data transferred after the last checkpoint is resent.

It also manages the 'dialogue' between the computers on the network. An example of this is half duplex mode, this is when it is responsible for determining whose turn it is to transmit over the network.

This layer communicates through gateways and application interfaces and uses services such as TCP.

The protocols associated with this layer include:
▶ NFS
▶ SQL (Structured Query Language)
▶ RPC (Remote Procedure Cell).

Transport layer

This layer is responsible for ensuring reliable data delivery so that packets (formatted units of data) arrive error free and without loss. It uses messages that inform the sender that the data was successfully received. If data is not delivered, then the message received from the sender will result in a retransmission of the data. If the data received is in a damaged state, the message known as NACK (negative acknowledgement) is sent and retransmission is forced.

The transport layer provides a service for connection-mode transmissions and connectionless-mode transmissions. With connection-mode transmissions, a message is sent or received in packets which then need to be reconstructed into a complete message.

This layer communicates through gateway services, routers and **brouters**.

> ### Key term
>
> **Brouters:** these function as both bridges and routers.

The protocols associated with this layer include:
▶ TCP
▶ UDP (User Datagram Protocol)
▶ SPX (Sequenced Packet Exchange)
▶ NetBEUI (NetBIOS Extended User Interface).

Network layer

This layer is responsible for moving data around a network of networks, commonly known as the internet. It transfers information between networks by examining the logical network address and routing the packets using routers. The path or route taken to the destination network address is determined either statically or dynamically. The packet moves one step at a time through the internet to the target network. The hardware address is then used to move the packet to the target node. This process requires each logically separate network to have a unique network address. The Internet Protocol (IP) addresses make it easier to set up a network and connect to other networks.

In order to make it easier to manage the network and control the flow of packets, the network layer addressing is separated into smaller parts known as subnets. It is the subnet portion of the IP addressing that is used to route traffic between different networks. Routers must be configured specifically for the networks or subnets that will be connected to them.

Other functions carried out by the network layer include:
▶ Error control – detection of transmission errors and retransmission of correct data.
▶ Flow control – regulating the speed of data transfer.
▶ Fragmenting packets – breaking down packets into smaller chunks if required. The receiving network layer is responsible for rebuilding the packets.

This layer communicates through gateway services, routers and brouters.

The protocols commonly associated with this layer include:
▶ IP
▶ IPX (IP Exchange)
▶ RIP (Routing Information Protocol)
▶ ARP (Address Resolution Protocol)
▶ ICMP (Internet Control Message Protocol)
▶ RARP (Reverse Address Resolution Protocol)
▶ EGP (Exterior Gateway Protocol)
▶ NetBEUI
▶ DLC (Data Link Control).

Data link layer

The data link layer is responsible for transferring data between devices. It responds to requests from the network layer above it and issues requests to the physical layer below it.

It is responsible for:
▶ encoding bits into packets prior to submission and decoding the packets back into bits at the destination
▶ the logical link control (LLC), media access control (MAC), hardware addressing, error detection and handling.

The data link layer is divided into two sublayers: LLC and MAC. The former controls how computers on the network gain access to the data and obtain permission to transmit it; the latter controls packet synchronisation, flow control and error checking.

Data link layer processing is faster than network layer processing because less analysis of the packet is required.

This layer communicates through switches, bridges and intelligent hubs.

The protocols associated with this layer include:
- HDLC (High-level Data Link Control)
- LLC
- SLIP (Serial Line Internet Protocol)
- PPP (Point-to-Point Protocol).

Physical layer

This is responsible for the transmission of data over network communication media. The physical layer includes:
- the network medium
- physical network topologies
- the network card
- the process of transmitting and receiving signals from the network medium including bit transmission, encoding and timing rules.

The four general functions of this layer are:
- Definitions of hardware specifications – each piece of hardware on a network will have a specification, for example the maximum length of cable, EMI protection or width of cable.
- Data transmission and reception – regardless of the network medium used, there has to be equipment that actually transmits the signal and equipment that receives the signal; for example optical transmission lines use equipment which can produce and receive pulses of light, such as amplifiers and repeaters.
- Encoding and signalling – this is a very important part of the physical layer and can be quite complicated.
- Topology and physical network design – the physical layout and structure of the network.

This layer communicates through repeaters, hubs, switches, cables, connectors, transmitters, receivers and multiplexers.

Table 7.10 provides a summary of the layers, their purpose and their location within the OSI seven-layer model.

Layer no.	Layer title	Purpose	Location
7	Application Supports the applications and end user processes. It is not an application itself.	Messages and packets • Identification of communicators (Is there anybody out there?) • Assessment of network capacity (Will the network let me contact them now?) • Data syntax (Can we understand the message?) • User authentication and privacy (How do you know it is me?)	Upper/host Application-specific function: • Formatting • Encryption • Connection management
6	Presentation Usually part of the operating system	Packets Converts data into a suitable format so it can be understood by the application. It also: • supports data compression and encryption • decides data structure • communicates through gateways and application interfaces.	
5	Session Allows applications running on different computers to communicate with each other	Packets • Sets up the communication link between applications • Manages the link, e.g. if using half duplex, this layer determines whose turn it is to transmit data • Terminates the link • Authenticates and reconnects the link after an interruption	

Layer no.	Layer title	Purpose	Location
4	Transport The postal service, ensuring that data packets arrive error free and without loss	Datagrams, segments and packets • Puts the data into the correct packet format • Delivers the packet • Checks the packet has arrived • Retransmits the packet if not received • If data packet is damaged, arranges for resubmission	Lower/media Provides: • Routing • Addressing • Flow control
3	Network Transfers data around between the internet	Datagrams and packets • Examines the logical network address • Converts it into physical machine addresses on the receiving computer and reverses it when a message is sent back • Routes packet using routers • Detects transmission errors (error control) • Retransmits correct data • Regulates the speed of data transfer (flow control) • If the packet is too large for a network on the route to handle, it is fragmented and reassembled by the receiving device (fragmenting packets).	
2	Datalink Responsible for transferring data between devices. It responds to requests from the network layer above and the issues requests to the physical layer below.	Bits and packets • Encodes bits into packets prior to submission • Decodes packets back into bits Divided into two sublayers: • Media access control (MAC) layer which controls how computers on the network gain access to the data and object permission to transmit it • Logical link control (LLC) layer controls packet synchronisation, flow control and error checking	
1	Physical Includes: • Network medium • Physical network topologies • Network card • Process of transmitting and receiving signals from the network medium	Four generation functions: • Definition of the hardware specification, e.g. maximum length of cable, width of cable, physical connectors, voltages • Data transmission and reception, e.g. amplifiers and repeaters • Encoding and signalling • Topology and physical network design, physical layout and structure of network	

▲ Table 7.10 A summary of the layers, their purpose and their location within the OSI seven-layer model

Activity

Complete the table by providing the full name of the protocols relevant to each of the OSI layers and a brief description of their role. The first part is completed for you as an example.

OSI layer	Protocol	Full protocol name	Description
Application	FTP	File transfer protocol	The simplest method for sending and receiving files over the internet, FTP splits the files into a number of segments and gives each one a reference number so that the …
	TELNET		
	WWW		
	HTTP		
	TCP		
	NFS		
	SMTP		
Presentation	JPEG		
	MIDI		
	MPEG		
	Music, images, movie formats		
Session	NFS		
	SQL		
	RPC		
Transport	TCP		
	UPD		
	SPX		
	NetBEUI		
Network	IP		
	IPX		
	RIP		
	ARP		
	ICMP		
	RARP		
	EGP		
	NetBEUI		
	DLC		
Datalink	HDLC		
	LLC		
	SLIP		
	PPP		
Physical			

How data is transferred between network devices using the OSI seven-layer model

▶ Data is transmitted by the user from the digital device, for example computer to the application layer.

▶ Data is then passed down through the layers to the physical layer. This is the only layer that is able to communicate with other devices and networks.

▶ The data is sent along the physical layer to the receiving device, for example receiving computer.

▶ The data then works its way up through the layers to the receiving computer for access by the user receiving the data.

Test yourself

1 Identify the layer of the OSI model which is sometimes referred to as the protocol layer.
2 Explain the function of the presentation layer of the OSI model.
3 Describe the physical layer of the OSI model.
4 Describe the datalink layer of the OSI model.
5 Where does the transport layer sit in the OSI model and what is its role?

Application layer

This provides applications with the means to access the services of the other layers and defines the protocols used by the applications to exchange data. The most widely known protocols used for the exchange of user information are:

▶ HTTP – Hypertext Transfer Protocol (for example web pages)

▶ FTP – File Transfer Protocol (for example interactive file transfer)

▶ TELNET – used for logging on to networks remotely

▶ SMTP – Simple Mail Transfer Protocol (for example transfer of email messages and any attachments).

Key protocols are used to assist in the management and use of TCP/IP networks:

▶ DNS – Domain Name System (for example linking a host name to an IP address).

▶ RIP – Routing Information Protocol (used, for example, by routers to exchange routing information on an IP network).

▶ SNMP – Simple Network Management Protocol (used, for example, to collect and exchange network management information).

Transport layer

This layer is responsible for providing the application layer with session and datagram communication services. The main protocols are:

▶ TCP (Transmission Control Protocol) – This provides a one-to-one communication services and is responsible for the:
  – establishment of the TCP connection
  – sequences and acknowledgement of packets sent
  – recovery of packets lost during transmission.

▶ UDP (User Datagram Protocol) – This provides a one-to-one or one-to-many communication service which is connectionless and unreliable. It is used when the amount of data to be transferred would fit into a single packet. It is used by network applications that want to save processing time.

Internet layer

This layer is responsible for the addressing, packaging and routing functions. The main protocols are:

▶ IP (Internet Protocol) – Responsible for the IP addressing, routing, fragmentation and re-assembly of packets.

▶ ARP (Address Resolution Protocol) – It ensures that the address of the internet layer can be linked to the network interface layer address, for example the hardware address.

▶ ICMP (Internet Control Message Protocol) – This provides the diagnostic functions and reporting errors when delivery of IP packets is unsuccessful.

▶ IGMP (Internet Group Management Protocol) – This is responsible for the management of IP groups.

Network interface layer

This layer is responsible for placing TCP/IP packets on the network medium and receiving them off the network medium. TCP/IP can be used to connect different network types, for example LAN technologies such as token ring and WAN technologies such as frame relay.

7.2.10 The four-layer TCP/IP model to describe how applications communicate over a network, including the function and related protocols of each layer

Layer no.	Layer title	Purpose
1	Application	Provides applications with the means to access the services of the other layers and defines the protocols used by the applications to exchange data.
2	Host–host/transport	Provides the application layer with session and datagram communication services.
3	Internet	Provides addressing, packaging and routing functions.
4	Network access	Physical interface between the host system and the network hardware.

▲ Table 7.11 TCP/IP layers

Transmission Control Protocol/Internet Protocol (TCP/IP) is the protocol used for communication between computers on the internet. It defines how devices should be connected to the internet and how data is transmitted between them.

There are four layers to TCP/IP:
▶ application layer
▶ transport layer
▶ internet layer
▶ network interface layer.

7.2.11 Understand the role of data packets in transmitting over a network

Data that is sent over a network (including the internet) is referred to as packets. Large packets of data are broken down into smaller datagrams. On reaching their final destination, the datagrams are reassembled as a single file or block of **contiguous data**. The term 'packet' and 'datagram' are similar in meaning. The protocol UDP uses the term 'datagram'.

Data packets consists of a header, a payload and a trailer.
▶ Header – this contains the instructions about the data and has several parts:
  – originating address – the IP address of the sender of the data packet

– destination address – the IP address of the receiver of the data packet
– internet protocol – defining the type of packet being transmitted, for example email, webpage, video, etc.
– the size of the header and the payload
– the number of hops – this is the number of routers that the packet will pass through on its journey
– Time to Live (TTL) – the amount of time it exists within the network before being discarded by the router
– flags – used to inform the router whether the packet can be divided into datagrams
– checksum – used to detect any errors during transmission
– packet number – the number of the packet where there is a sequence of packets.

Key terms

Contiguous data: data that is stored in a collection of adjacent locations.

Time to live (TTL): the amount of time or 'hops' that a packet is set to exist inside a network before being discarded by the router. TTL is also used in Content Delivery Network (CDN) caching and Domain Name System (DNS) caching.

Hops: refers to the number of routers that a packet passes through from its source to its destination. A hop can also be counted when a packet passes through other hardware on a network such as switches, access points and repeaters. It is dependent on what role the devices have on the network and their configuration.

▶ Payload – this is the body of the data packet and contains the actual data that is being sent/received.
▶ Trailer – this is sometimes referred to as the footer and is used to inform the receiving device that it is the end of the packet. The trailer also includes error checking, the most commonly used is the Cyclic Redundancy Check (CRC). The CRC will add up all the 1s in the payload and stores the results as a hexadecimal. On receiving the data packet, the receiving device will add up the number of 1s in the payload and compare it with the hexadecimal value stored in the trailer. When the two values match it is confirmation that there has been no error during transmission. If, however, the two values do not match, then the receiving device sends a request to the sending device asking it to resend the packet.

Packet switching

This is where the data packet is divided into smaller data packets and transmitted individually across the network as opposed to transmitting one large data packet. It is used to reduce the chances of lost packets and facilitates the resending of data packets as well as reducing transmission latency. The packets are not necessarily routed along the same path within the network. This results in the packets arriving at their destination in no particular order. It is the responsibility of the destination to reconstruct the packets into an appropriate order to be able to retrieve the original message.

Error handling – CRC

This was discussed under the section 'Trailer'.

Test yourself

1 Explain the error checking method CRC.
2 What are large packets divided into?
3 How many components has a packet?
4 Describe the payload component of a packet.
5 Explain the term 'packet switching'.

Layer	Protocol suite					
Application	Telnet	FTP	SMTP	DNS	RIP	NMP
Host–host/ transport	TCP			UPD		
Internet	IP				IGMP	ICMP
	ARP					
Network access	Ethernet	Token ring		Frame relay	ATM	

▲ Table 7.12 TCP/IP layers and associated protocols More detail about protocols is in section 7.2.12.

Table 7.13 shows the important differences between the OSI and TCP/IP models:

TCP/IP model	OSI model
Has four layers.	Has seven layers.
Uses a horizontal approach.	Uses a vertical approach.
Protocol orientated approach.	Based on the functionalities of the layers.
Combines the presentation and session layers into its application layer (so the application layer contains the application, presentation and session layers that are separate layers in the OSI model).	Differentiates between interfaces, services and protocols.
Combines the data link and physical layers and is referred to as the network layer (sometimes referred to as Network Access Layer).	Acts as an interaction gateway between the network and end-user.
Only connectionless transmission is available in the network layer. Connection and connectionless transmission is available in the transport layer.	In the network layer, connection and connectionless transmissions are provided. It only provides connection transmission in the transport layer.

▲ Table 7.13 The differences between OSI and TCI/IP models

Test yourself

1 Identify the layer in the TCP/IP model which functions as three layers of the OSI model.
2 Explain the purpose of the host–host layer.
3 The network access layer in the TCP/IP model functions the same as the data-link layer and which other layer in the OSI model?
4 Discuss the important differences between the OSI and TCP/IP models.
5 Identify three protocols in the internet layer.

7.2.12 The role of common network protocols

Network protocol	Function
TCP/IP (Transmission Control Protocol/Internet Protocol)	A set of standardised rules that allow computers to communicate on a network (such as the internet)
DNS (Domain Name System)	When accessing a website a user can enter the IP address or use the hostname address. But IP addresses are not as easy to memorise. The DNS converts the IP address into a host name that can be read easily by the users and remembered by them for future reference.
ARP (Address Resolution Protocol)	This is a communication layer protocol that supports the process between the datalink and network layers. It identifies the designated MAC address based on the available IP address.
RDP (Remote Desktop Protocol)	This is a Microsoft product that provides users with a graphical user interface (GUI) when connecting to other networked computers.
VNC (Virtual Network Computing)	This is used for remote desktop sharing and displays the virtual desktop of another computer, allowing it to be controlled over a network connection.
SSH (Secure SHell)	Used to securely operate services over an unsecured network using cryptography. It is commonly used by network administrators who are required to access, manage and control systems and applications remotely.
SIP (Session Initiation Protocol)	This is known as a signalling protocol that is able to initiate, maintain, adjust and terminate real-time sessions. Examples include voice and video messaging between connections on an IP network.
SMB (Server Message Block)	This is a network communication protocol that enables the shared access of files, printers and ports between connected nodes on a network.
FTP/S (File Transfer Protocol/Secure)	This is based on the client and server model architecture and is used to transfer files between the client and the server on the network.
HTTP/S (Hyper Text Transfer Protocol/Secure)	HTTPS is used to authenticate a website that the user is attempting to access. It provides secure network communication through the protection of the privacy and integrity of any data that is exchanged. This is achieved through the encryption of the exchanged data and is used, for example, for banking and e-commerce websites. HTTP provides communication but there is no securing of data. HTTP is commonly used for internet forums and websites where the exchange of sensitive data is not required.
IMAP (Internet Message Access Protocol)	This is an internet email protocol that stores emails on the mail server. The users then can retrieve the messages and respond to them.
POP3 (Post Office Protocol)	This is within the application layer internet protocol and facilitates the retrieval of emails from a remote server to the user's local device. This enables the user to view emails offline.
SMTP (Simple Mail Transfer Protocol)	This is a communication layer protocol that is used to send emails.
SNMP (Simple Network Management Protocol)	This collects and organises information associated with the devices that are managed on an IP network. It can modify the information in order to change the behaviour of the connected device. Devices include routers, switches and printers.
TELNET	This is referred to an application protocol which allows for bidirectional interactive communication using a virtual terminal connection.

▲ Table 7.14 Network protocols and their functions

Test yourself

1 Describe the role of IMAP.
2 What does the acronym RDP stand for?
3 What is the role of the Secure SHell protocol?
4 Which protocol provides bidirectional interactive text-orientated communication using a virtual terminal connection?
5 Identify an internal email protocol and describe its role.

7.2.13 The use of physical, virtual and cloud environments to solve problems and meet the needs of organisations and their stakeholders

In sections 3.2.1 and 6.1.3 you learned about the IoT, and how it is used by organisations to carry out the business functions and meet the needs of their stakeholders. In section 7.3 you will learn about virtual environments, and in 7.4 you will learn about cloud environments. You will be able to understand how a combination of one or more of these concepts can be used by organisations to solve problems, improve business processes and meet the needs of the stakeholders.

7.3 Virtual environments

7.3.1 The key features of virtual environments

Virtualisation is technology that enables a single, physical hardware system to be separated into multiple simulated environments or dedicated resources. Software known as a **hypervisor** connects to the hardware and facilitates the splitting of one system into separate and distinct environments referred to as VMs. The VMs rely on the hypervisor to separate the resources of the machine from the hardware and distribute them. The hardware that is equipped with a hypervisor is called the host and the VMs are called guests. The VMs (guests) treat the resources such as CPU, memory and storage as a pool of resources. These resources are controlled by operators so that the VMs receive only the resources they need and when they need them.

Examples of types of virtualisation

▶ **Data virtualisation** – enables organisations to treat data as a dynamic supply. This provides processing functions that can bring data together from multiple sources, transform data and accommodate new data sources.
▶ **Desktop virtualisation** – allows an organisation to install multiple OSs onto a single machine. It allows a central administrator (or an automated administrator tool) to install simulated desktop environments on very large quantities of physical machines simultaneously. Unlike the traditional desktop environments, they are physically installed, configured and updated on each machine. Desktop virtualisation allows the administrators to carry out mass configurations, updates and security checks on all virtual desktops.
▶ **Server virtualisation** – you previously learned that servers are computers which are designed and configured to process high volumes of specific tasks. Virtualising a server provides it with the ability to perform even more functions and involves partitioning. This enables it to serve multiple functions.
▶ **OS virtualisation** – this allows an organisation to run multiple OSs side by side, for example Windows and Linux, making different OSs accessible to different computers. This reduces the hardware costs and increases security as all virtual instances can be isolated and monitored. It also reduces the time required for updates to be installed.
▶ **Network function virtualisation** – separates the network's key functions, for example file sharing and IP configuration, so that they can be distributed among different environments. When software functions are independent of the physical machines, functions can be packaged together to form a new network and assigned to a particular environment. It reduces the number of physical components required to create multiple and independent networks.

Key features

▶ **Increased security** – a VM manager can control and filter the activity of a guest's programs. This can mitigate the risk of harmful operations being carried out. Any resources exposed by the host can be hidden or protected from the guest.
▶ **Managed execution** – the execution of sharing, aggregation, emulation and isolation can all be managed from a central virtual server.
▶ **Sharing** – this is the creation of separate computing environments within the same host. It is used to reduce the number of active servers and limits power consumption.
▶ **Aggregation** – as well as sharing physical resources among numerous guests, virtualisation allows aggregation. This is where a group of hosts are combined and presented to guests as a single virtual host. This is made possible through cluster management software.
▶ **Emulation** – guest programs are executed in an environment that is controlled by a virtualisation layer. This is basically a program. In addition, a

totally different environment with respect to the host can be emulated. This allows the execution of guest programs that require specific characteristics which are not available on the physical host machine.

▶ **Isolation** – virtualisation provides guests with a separate environment, regardless of whether these are OSs, applications or other entities. The guest program interacts with an abstraction layer which provides access to the underlying resources. This allows the VM to filter the activities of the guests and mitigate the risk of harmful operations against the host. It also facilitates performance tuning, where the performance of a guest can be monitored and controlled through the fine tuning of the properties of the resources available through the virtual environment. It therefore supports and implements a quality-of-service (QoS) infrastructure.

▶ **Portability** – the portability differs with specific types of virtualisation.
  – With a hardware virtualisation environment, the guest is packaged into a virtual image. This means that it can be safely moved and executed on top of different VMs.
  – With programming-level virtualisation, the binary code representing application components can run without any recompilation on any installation of the corresponding VM.

Test yourself

1 Explain the term 'virtualisation'.
2 What are the benefits of desktop virtualisation?
3 Aggregation is a feature of virtual environments. What does this mean?
4 Describe data virtualisation.
5 Explain how a virtual environment can improve the security of an organisation's digital systems.

7.3.2 The benefits and drawbacks of the use of virtual environments for organisations in a variety of contexts

Benefits of using virtual environments

▶ **Efficient use of hardware** – many organisations spend a lot of money implementing their digital systems and servers. Unfortunately, many organisations only use a fraction of them effectively. With virtualisation, multiple instances of the same hardware can be created and therefore maximum effective use can be made of them by a wider variety of guests. So, costs can be reduced and effectiveness increased.

▶ **Continuous availability** – virtualisation allows access at any time. A virtual instance can be moved from one server location to another, without having to shut down and restart the processes already in operation. Also, data is not lost during any migration process. Therefore, data is unaffected by any unplanned **outages** as the system will always be running.

▶ **Simple recovery** – due to the virtual instances on remote servers, duplication, backups and recoveries are much easier. Tools are now available that provide close to real-time data backup and mirroring. This results in zero data loss at any given point in time. If a system crashes or there is some form of downtime, the last saved position mirrored on another virtual instance can be accessed and run. Organisations can attain a high level of efficiency because business can continue.

▶ **Simplified and faster setup** – setting up systems and servers can be complex and time-consuming. Systems must be physically connected and then OSs and other software installed. This can take many hours to complete. With virtualisation, an organisation can be up and running within minutes.

▶ **Easier migration to the cloud** – the challenge for any organisation when migrating to the cloud, is the migration of the vast quantities of data available on their premises. This is easier to carry out with virtualisation because most of the data is available on a server.

Key term

Outage: a period when the power, a service or equipment is closed down.

Drawbacks of using virtual environments

▶ **High initial investment costs** – while virtualisation helps organisations reduce operational costs, the initial setup costs of servers and storage is higher than a regular installation of a system.

▶ **Risk to data** – working on virtual instances on shared hardware resources means that an organisation's data is hosted on a third-party resource. These means that the data is vulnerable to attacks and/or unauthorised access. This can be a huge challenge if the service provider does not have robust security solutions to safeguard the organisation's virtual system and data. This is particularly apparent with storage virtualisation.

▶ **Scalability is challenging** – while scaling on virtualisation is simple, it is not so easy to implement in a short period of time. With a physical setup, new/additional hardware is relatively straight forward even if there are a few minor complications during installation. With virtualisation, the software, security, storage and resource availability have to be considered carefully and factored in. This can be time-consuming because a third-party provider is involved.

▶ **Reduced performance** – virtualisation facilitates the optimum use of all resources, but it is also challenging when the system requires additional performance such as speed and it is not available. Because the resources are shared, what may have been accessed previously by a single user is now shared with multiple users. These resources may not be shared equally or may be shared depending on the task being carried out. The greater the complexity of the tasks, the greater the need for improved performance from the system. If the performance is not available, then the time taken to complete a task increases.

▶ **Unintended server sprawl** – this is a major concern for server administrative staff and users. Many issues that are raised are in relation to server sprawls. The setting up of a physical server is time-consuming and consumes resources. A virtual server can be created in minutes. Instead of using the same virtual server, users create new servers allowing them to 'start again from fresh'. This means that a server administrator that would normally have four, five or six servers, now has to handle 20 or more virtual servers. This complicates smooth operations of systems and if the termination of a server is forced, it can also mean the loss of data.

Test yourself

1 Discuss the benefits to an organisation of using virtual environments.
2 Describe the term 'outage'.
3 Explain why a virtual environment can reduce performance.
4 Identify one reason why there is a higher security risk to data when using virtual environments.
5 Describe the term 'unintended server sprawl'.

Activity

You have been asked to put together an information guide for an organisation that is considering implementing a virtual environment. They do not know anything about virtual environments, just that they can be good. In the information guide, you are required to explain what a virtual environment is, the different types of virtualisation and the key features. The organisation would also like to know what the benefits and drawbacks are of having a virtual environment.

7.4 Cloud environments

7.4.1 The concepts of cloud computing deployment

7.4.2 Common cloud delivery models and the way in which responsibility and ownership of resources are distributed between the subscriber and the service provider

These two sections have been combined because it is important to understand the delivery models and how they have an impact on the deployment of cloud computing.

Cloud computing refers to the outsourcing of IT services and infrastructure, making them accessible remotely via the internet. Utilising cloud computing models can boost productivity and provide organisations with a competitive edge. There are a number of different types of cloud delivery models providing a variety of cloud solutions. Which is chosen will depend on what an organisation wants to outsource.

There are four different categories of cloud models. These are:

▶ Public – these are used by a range of end-users, from large organisations to individuals. They can be accessed through a digital device with an internet connection. To be able to access and use a public cloud, users must pay either a subscription or on a pay-to-use basis. A public cloud is usually hosted by one cloud vendor. Examples of public clouds include Amazon Web Service (AWS), IBM Cloud or Google Cloud.

▶ Private – these are used by a specific organisation and can be hosted on site premises or remotely through virtualisation. Access to a private cloud is restricted by a firewall which is managed by the organisation. A private cloud requires that the organisation installs and maintains infrastructure, including hardware, software and networking components.

▶ Community – a community cloud is one that is designed for and used by a group of organisations working in a particular industry, for example the financial sector. By using a community cloud the financial sector can transfer money between banks, process credit and debit card payments and liaise with international banks. An example of a community cloud is the SWIFT banking system.

▶ Hybrid – the hybrid model is the 'best of both worlds' and incorporates the best aspects of private and public clouds. For example, an organisation can utilise some services on the public cloud but has a private cloud for sensitive services or data.

Delivery models of cloud computing

There are four commonly used cloud delivery models:
▶ Infrastructure as a Service (IaaS)
▶ Platform as a Service (PaaS)
▶ Software as a Service (SaaS)
▶ Data as a Service (DaaS).

> DaaS is covered in section 7.4.3.

Each delivery model has a different purpose. More than one delivery model can be used by an organisation to provide a complete cloud service which meets their needs.

IaaS provides organisations with infrastructure. The infrastructure provided by the IaaS provider can include a network, servers, storage areas and an operating system (OS). This means that the organisation can reduce the costs and physical presence of hardware. One of the benefits of using an IaaS is that organisations only pay for what they use and so can increase, or decrease, the infrastructure they use. This scalability means that the organisation can be flexible as its requirements change. Examples of IaaS include Amazon Web Services (AWS), Microsoft Azure and Oracle Cloud.

PaaS includes all the features of IaaS and builds on these to include software such as development tools. But it is still possible for the organisation to use their own application software. PaaS is commonly used during the development and testing of a (usually) web-based software application. The PaaS provider provides the platform, hardware and software to develop the software application on, and the organisation remains responsible for developing and deploying the application. As with the IaaS model, PaaS is scalable. Examples of PaaS include Google App Engine and Microsoft Azure.

SaaS is a very popular delivery model with organisations and, increasingly, individuals. SaaS provides the software applications and services, for example a cloud storage area, that the end-users can access. As the software provided by SaaS is not location dependant, software can be accessed from a range of devices in different locations as long as an internet connection is available. The software is accessed through a subscription service; this may be a fixed fee with limitations, such as the number of devices that can access the software, or on a pay-to-use basis. Examples of SaaS include Microsoft 365 and Google Docs.

Each delivery model differs in the way in which responsibility and ownership of resources are distributed between the subscriber and service provider. Table 7.15 shows where the different responsibility and ownership lies for each of the cloud delivery models.

	Subscriber (organisation)	Service provider
Iaas	• Applications • **Runtime** • Data • Middleware • OSs	• Virtualisation • Servers • Storage • Networking
PaaS	• Applications • Data	• Runtime • Middleware • OS • Virtualisation • Servers • Storage • Networking
SaaS	• User only	• Applications • Data • Runtime • Middleware • OSs • Virtualisation • Servers • Storage • Networking

▲ Table 7.15 Cloud models: IaaS, PaaS, SaaS

Key term

Runtime: the length of time a computer program takes to run.

7.4.3 The concept of DaaS (Data as a Service)

DaaS is a cloud computing model that provides data management by using the cloud for the storage, integration, processing and, if required, analytics of data through a network connection. DaaS is facilitated through SaaS. Organisations will use DaaS to control the storage, processing, etc. of data using the cloud, which reduces **data sprawl** and **data silos**. This means that DaaS is useful for sharing data between teams and encouraging greater collaboration between different departments.

Data science platforms

Data science platforms deal with the analysis and accessibility of data, and therefore benefit from using DaaS. It is software that provides data scientists with the tools to effectively collaborate on work products while working through the data science lifecycle. The data science platforms support data scientists when they are exploring and developing ideas and models. This involves qualitative research and can involve plotting vast amounts of data. A data science platform provides the tools to allow the team to collaborate and identify data that has already been gathered (and possibly analysed), explore vast amounts of data and facilitate the opportunity to use new tools and techniques without disrupting work that has already been carried out. During this phase, data scientists will be able to develop ideas and run experimental models that they can share with their colleagues.

Dashboards

These are commonly referred to as **data visualisation** tools as they enable the user to view a visual representation of the data. As dashboards are heavily involved with data analytics, such as providing a

visual interpretation of **performance metrics**, the DaaS infrastructure is an appropriate platform for it to use for the accessibility of data, irrespective of its geographical location.

Refer also to section 6.3.2, in Content area 6.

Business information tools

This was covered in section 6.3.2, in Content area 6.

Data lakes

This was covered in section 6.4.2, in Content area 6.

Key terms

Data sprawl: the vast amounts and variety of data produced by organisations on a daily basis.

Data silos: a group of raw data accessible by one department but not available to the other departments within the organisation.

Data visualisation: the graphical representation of data.

Performance metrics: the collection, analysing and reporting of information. This can be in relation to a system, department, individual or even profit of an organisation as a whole.

Databases

Databases can store vast amounts of data and therefore the implementation of a DaaS facilitates the accessibility of the data and information by a wide range of people who may be based in the same or different locations. Using DaaS also enables an organisation to store big data without the necessity to install its own data storage servers.

File systems

DaaS supports the sharing of files as the data is stored in the cloud and facilitates simultaneous access to multiple users to authorised files on the cloud. The authorised users are able to create, modify, read, write and delete files. Users are granted permissions through the implementation of user and group permissions which strengthens the security of the files. These permissions are controlled and managed by the administrator.

7.4.4 How DaaS is used by organisations, and the benefits and drawbacks it provides to organisations and their stakeholders

As discussed in section 7.4.3, DaaS is used by organisations for the storing, processing, analysis and sharing of data and files across different users, teams and departments regardless of the location of the individuals.

Benefits	Drawbacks
Cross platform transfer (transferring data between different systems)	Data is transferred over a network, increasing the risk of security breaches
Automatic updating of data/files	Time consuming for the transferring of large data sets based on bandwidth restrictions
Wide ranging accessibility, e.g. global	Can be a limitation of available tools as any tool used must be hosted by or compatible with the DaaS platform
Reduces the risk of multiple versions of the same data/files existing	Compliance with legislation/regulations as sensitive data is located in the cloud. Organisations must ensure that they are compliant with the regulation/legislation associated with the country where the DaaS provider is based and the origin of the data
Enhanced security through access permission levels, passwords and encryption	
Reduction in the costs of data maintenance and delivery	
Facilitates collaboration, regardless of location	
Allows for easier scalability (up and down) for storage requirements	
Supports remote working	

▲ Table 7.16 Benefits and drawbacks of DaaS for organisations and stakeholders

Test yourself

1 Discuss the benefits and drawbacks to a scientific research team who are considering using DaaS.
2 Identify two barriers that organisations may have when trying to carry out big data analytics.
3 Explain one benefit and one drawback to a retail organisation who are considering storing the information about their stock and customers on a DaaS platform.
4 Describe how DaaS prevents the storage of numerous versions of data in different locations.
5 Why can organisations have compliance issues if they deploy a DaaS platform?

7.4.5 Cloud sourcing and cloud portability, and their implications for service providers and organisations (subscribers)

Cloud sourcing

This is where organisations (known as the subscribers) will outsource business processes to a third party (referred to as a service provider). Subscribers pay the service providers for services such as IaaS, PaaS, SaaS and DaaS. Dropbox is a well-known cloud source used by individuals for the storage and sharing of data and files.

Cloud sourcing provides subscribers with an easier option for scalability (whether this is upward or downward). Subscribers only need to contact their providers and renegotiate the fee in order for their needs to be met. In addition, the onus is on the service provider to ensure that the services they offer are up-to-date and maintained as opposed to subscribers having to employ technical teams to maintain these services for them.

The market for cloud sourcing has expanded rapidly from small service providers to much larger ones. The smaller service providers have to consider new and innovative ways to retain their share of the market.

Cloud portability

This is the ability for organisations to move their applications, platforms and/or data between service providers. Cloud portability can be very complex.

Research the impacts on organisations and providers with respect to cloud sourcing and cloud portability.

Using the table, enter the results of your research in a similar format to the entry made for data security.

Impact	Subscribers (organisations)	Provider
Cloud sourcing – data security issues	Loss of customers, loss of business, loss of finance, potential legal implications	Providers need to instil confidence in the subscribers that their data will be secure. Failure to do this will result in a loss of business as subscribers will not purchase/rent their services.

Providers want to 'lock-in' their subscribers and therefore, often, do not offer portability between platforms. However, this can also mean that potential subscribers will not sign up to use their services because of the restrictions with portability.

Section 7.4 has looked at numerous examples of the benefits and drawbacks for organisations (the subscribers). The challenge to service providers is in relation to what they provide, how they provide it and how they protect a subscriber's data and remain compliant with legislation. Organisations must consider what would be the most appropriate form of cloud computing for the business based on their needs and those of other stakeholders.

7.5 Resilience of environment

7.5.1 Ensuring digital environments are resilient and the impact on organisations and stakeholders if this is not achieved

Digital resilience involves:
▶ understanding the risks involved with having a digital environment and identifying how these risks can be mitigated
▶ being able to identify when a problem has occurred and what to do

▶ being able to recover from problems
▶ learning from experience and implementing processes and procedures to mitigate any future occurrences of the same or similar problems.

Strong digital resilience is when organisations manage digital risk and, at the same time, continue to deliver their products/services to stakeholders regardless of the situation. It is a combination of business resilience and cyber security.

Business processes are becoming more and more reliant on technology, and should a problem occur that disrupts the functionality of the business, it can result in the loss of customers, revenue and reputation, as well as important/sensitive business data. It is therefore important that businesses have plans in place that enables them to adapt to these problems, allowing them to function with minimal disruption to the business and to their stakeholders. It also involves businesses considering new technology and embracing the opportunities that this technology provides, to ensure that they remain competitive within the marketplace.

When considering digital resilience, organisations must be able to balance the opportunities for technological change, business expansion and security while complying with current legislation and regulation.

7.5.2 Methods used to improve the resilience of digital environments

7.5.3 The benefits and drawbacks of methods used to improve the resilience of digital environments

Data redundancy is where the same data is stored in multiple locations (a common practice in businesses). While data redundancy can occur by accident it is more often used for backup purposes to allow for data recovery should an incident occur. Another advantage is that it provides a mechanism for improving the reliability of the data. This is achieved by comparing the data stored in the different locations for completeness and accuracy. However, there are also disadvantages. Data redundancy can cause data inconsistency. For example, a customer database is stored in various formats in different locations/databases. If the email address is amended in one system but not the others, then the data becomes inconsistent. There is also the risk of data corruption

due to the transfer and/or storage of data between locations/systems. Obviously, there is the issue with additional costs required for the storage of the data and businesses need to consider whether the benefits of data redundancy outweigh the potential costs.

System redundancy can be in relation to hardware, software or both.
- ▶ Hardware redundancy consists of one or more additional complete hardware systems, for example additional servers that can be used should there be a failure with the hardware being used. So, if a server fails, then there is a backup server that can be used to allow the business to continue functioning. (Consider re-visiting section 7.1.5 relating to the use of RAIDs.) There are three types of hardware redundancy:
  - Dynamic – this is where only one system is working at a time. Should there be a failure the additional system is started so that the business processes can continue with minimum disruption.
  - Static – this is where systems run in parallel, all carrying out the same processes. The results of these processes are continually compared and should an issue arise, a predefined algorithm is used to select the most appropriate outputs.
  - Hybrid – this is a combination of dynamic and static hardware redundancy.
- ▶ Hardware redundancy does have the drawback of resulting in a more complicated system and additional costs for the components involved. These redundant systems also have to be updated, upgraded and maintained alongside the main system, otherwise it will become unreliable and less resilient.
- ▶ Software redundancy is where there is additional software that can carry out the same tasks as the current software being used. As with hardware redundancy, it can be dynamic or static and works in a similar way. It is also used to protect against specification/design problems that can occur with any software that is developed.

Back-up systems

A system backup is not just about backing up the data but also backing up the entire system, including operating systems, software applications and configurations. If an incident occurs, this enables a business to restore its system to a position prior to that incident occurring. As with all backups, the state at which a system can be restored is dependent on when the last system backup took place, for example one week ago or one or more days ago. While a system backup can restore the state of the system relatively quickly, any changes made since that backup was created will have to be implemented again to ensure that the system is current.

Hot, cold and warm sites

There are three types of disaster recovery sites: hot, cold and warm.

▶ Hot site – this is usually hosted in a different location and runs in parallel to the normal system. It replicates the system (hardware, software, network and internet connectivity) within the business environment and are usually maintained by a third party at a cost to the business. Because they run in parallel with the current business environment, they can allow a business to promptly restore functionality should an incident occur. As these sites also contain a backup of the business data, it provides the business with immediate access to up-to-date data without having to source the location of any other data backups. Some businesses have their hot site in the cloud.

▶ Cold site – this is the cheapest option as it does not involve all the facilities that are needed for a hot site. A cold site is only used when a critical disaster has occurred. As it is only a form of data centre, it does not contain a replica of the computer equipment, etc. used within the business environment. Therefore, should an incident occur, technical expertise will be required to install and configure the computer equipment so that the business can continue to function. This, of course, is time consuming and can prevent a business from getting back up and running promptly.

▶ Warm site – this is a combination of a hot and cold site. While the digital equipment is readily available for use, the business data must be restored onto the system. It is not as cheap to use as a cold site but also not as expensive as a hot site.

So, to summarise, a hot site is a fully functioning, up-to-date replica of the business environment with respect to equipment and data. It is more expensive, usually located remotely and hosted by a third party. The benefit is that a business can be up and running very quickly after a major incident. A warm site contains the basic equipment used in the business environment but does not include any data. This data has to be restored on the warm site systems before the business can continue to function. Cold sites only have the data and none of the digital equipment used by the business environment. This has to be installed and configured by technical specialists and therefore takes longer for a business to start operating at normal capacity again.

Data backup and recovery procedures

It is important that, should an incident occur, an organisation is able to access up-to-date and reliable data in order to continue functioning. Even if a digital system can be recovered quickly, it can still create problems for an organisation if the data is unavailable or out-of-date.

There are four main types of backups: full, incremental, differential and mirror.

Full backup

A full backup is the backup of every file and folder stored on the system. These are more time consuming to complete and require sufficient space for all of the data. However, it is a faster method to use when there is a requirement to restore lost data. Because it includes all files and folders, it is more robust and reliable than other forms of backup. Full backups are particularly important if the data within an organisation changes significantly on a regular basis. It is storage hungry, and this has to be considered by a business. There is little point in carrying out regular full backups if the data does not change very often. It is also time consuming, and consideration has to be made as to how often the full backups should take place.

Incremental backup

After a full backup of the system is created, an incremental backup can be used. This is when only the data that has changed since the last backup has been created is backed up. For example, a person starts working on a report and saves it to the system. The report is not complete as they have further work to carry out. The incremental backup will back up a copy of the report. The following day, the person continues working on the report and finalises it. The incremental backup that takes place will create a backup of the report if it has changed since it backed it up originally. Incremental backups do not use as much storage space as a full backup but use additional resources so that it can compare the current state of the data with its previous backup and identify what data has changed and therefore what requires backing up. It is more time consuming to restore data from an incremental backup than a full backup. This is because it has to analyse the data within the backup to establish the timestamp for when the data changed. Therefore, several incremental backups may be required to restore the data.

Differential backup

The data initially undergoes a full backup procedure. It will then only backup the data that has changed since the last full backup. Take, for example, the report we discussed in the incremental backup section. This report will be continually backed up (even once it has been completed) until the next full backup takes place. Some people confuse differential and incremental backups. An incremental backup only backs up data (files and folders) that has changed since the last back up (regardless of the type of backup). A differential backup will continually back up any data (file/folder) that has changed since the last full backup. It is easier to recover the data as all that is required is the last full backup and the latest differential data backup. Differential backups require more storage than the other types of backups.

Mirror backup

A mirror backup (sometime referred to as a mirror image) is not only a backup of the data, but also of the entire system, including operating system, applications, configurations, preferences, booting procedures and hidden files. Consider a home computer and taking a mirror backup of the hard drive. If the hard drive fails, a new hard drive can be installed, and the mirror backup taken from the failed drive can be restored onto the new hard drive. Mirror backups are faster than other backup types and create

a 'clean' copy without old or outdated files and folders. Refer to section 7.1.5 where the use of mirroring is discussed for RAID 1 and RAID 10.

Device hardening

Device hardening is the methods or processes used to eradicate any means of attack. This can be achieved in many ways, for example:

- disabling unused network ports
- strict password management and file permissions
- using multi-factor authentication using hardware tokens as well as passwords on networks
- updating computer systems with security patches as they become available (but it is also important that the system is tested to ensure that the patch update has not created any functional issues)
- removing all non-essential services and programs – the bigger the 'surface area' of the system (by this we mean how large the system is and what it contains), the greater the opportunity for a potential hacker to access the system; by removing non-essential services and programs, the 'surface area' is reduced
- setting time limits on access, for example using a timeout system so that a person is automatically logged out of a system or application if there is no evidence of its use. This comes with its drawbacks as it can be frustrating for the user to be locked out after a short period of time because they have, for example. answered a phone call. This can slow down the workflow and, therefore, productivity.

Careful planning has to take place when implementing device hardening. The implications to workflow and productivity must be considered as well as the potential for reducing the attack surface areas of the system.

Test yourself

1 Explain the difference between resilience and redundancy and how they support system and data dependability.

2 Discuss the benefits and drawbacks of full backups.

3 Describe the benefits and drawbacks of a mirror backup.

4 Explain the term 'device hardening'.

5 Identify four ways that device hardening can be implemented.

Skills practice

A scientific research company based in the UK is going to work on a project where they will collaborate with other scientific research companies in Africa, Asia, Europe and the USA. The research companies are aware that they will need to consider the IT infrastructure that they must implement to facilitate:

► collaboration between the scientists in the different geographical locations
► accessibility, storage, analytical capability and security of the scientific data
► compatibility of platforms being used in the different geographical locations
► backup and recovery of the data should issues occur.

You have been asked to consider, make recommendations and justify your decisions with respect to:

► the types of servers required by the scientific companies
► any smart/internet enabled devices to help secure the data
► types of physical storage that may be required
► the type, model, topology and protocols required for the network
► the use of a virtual environment and the type(s) of virtualisation
► the use of cloud computing including which type and delivery model
► backup and recovery systems for the data.

Assessment practice

1 Identify and describe three characteristics that make a computer powerful and universally useful.
2 Explain how an incremental backup is created.
3 Explain the term DNS server.
4 Discuss hot, warm and cool sites and how they can be used by organisations.
5 Describe the term 'cloud sourcing'.
6 Discuss the benefits and drawbacks of DaaS.
7 Identify two smart/internet enabled devices used by organisations in the logistics industry.
8 Explain the term 'SaaS'.
9 Describe the term 'hybrid cloud'.
10 Explain the term 'cloud computing'.

Content area 8: Security

In this content area you will learn about the potential risks and threats to the digital systems used by organisations. You will apply your understanding of the implications of these to digital systems, as well as to organisations and their stakeholders. You will also learn about the relationships between the different aspects of the data and information that an organisation stores and uses, including confidentiality, integrity and availability (CIA).

Each of these risks or threats can be mitigated against to limit its impact and to reduce the threat of

it happening again. You will learn about a range of measures that can be used to do this. You will learn about several types of security. Cyber security is the most important type in relation to digital systems, data and information. Physical security can also be used to protect digital systems, data and information, including CCTV and access badges.

Learning outcomes

In this content area you will learn about:
8.1   Security risks
8.2   Threat mitigation

8.1 Security risks

8.1.1 Maintaining privacy and confidentiality of an organisation's information and that of stakeholders

Confidentiality relates to data, while privacy relates to the individual. In this context, an 'individual' can be a single person, a business or an organisation.

The GDPR is covered in Content area 4, section 4.1.2.

The CIA triad is covered in section 8.2.1 of this content area.

Typically, an organisation will store information about:
- ► employee salaries
- ► employee perks
- ► client lists
- ► trade secrets
- ► sales numbers
- ► customer information
- ► news about pending restructuring.

It is important that this information is kept confidential. Any breaches relating to the information can have a serious impact leading to the possible loss of clients or business. This in turn can lead to a downturn in the health of the organisation which may, ultimately, lead to the organisation's failure.

Employee salaries and perks

Salaries and perks should only be known by the employee and the HR department. It is important that this information is kept confidential as different employees carrying out the same task may be paid different salaries based on the number of years they have worked for the organisation, their experience and other factors such as qualifications and training courses attended. It is not acceptable to pay different salaries on the basis of gender or certain other protected characteristics (see Content area 4, section 4.1.4) as this would contravene the Equality Act.

Client lists and customer information

All organisations interact with people – clients and customers. Client lists and customer information are business-sensitive information that result from these interactions.

A client list may include individuals but also a named representative from a different organisation or business. Client lists show anyone who interacts with the organisation and they should not be accessed by employees unless absolutely necessary. Clients may interact with the organisation by using the services provided. For example, a client may use the services of an organisation that provides cloud-based storage facilities. Many organisations will have a client relationship team that looks after clients so this team will need access to this information.

Customer information usually relates to those who buy goods or services. The information held about customers will typically include personal details such as name and contact details but may also include order history.

If the privacy and confidentiality of client lists and customer information are not maintained, the organisation could lose clients and customers. People should expect that any organisation storing their personal data will keep it safe and secure to limit any breaches. The breach of personal data can impact on the organisation and also the people whose data has been leaked.

Sales numbers and trade secrets

Most organisations have stakeholders. Depending on the size and type of the organisation these may be shareholders – **external stakeholders**. Employees can also be classed as **internal stakeholders**. Some organisations may have a policy of keeping stakeholders informed about sales numbers as this may have a financial impact. Some organisations provide a financial bonus to employees or a dividend to investors or shareholders based on the previous year's sales numbers. Sales numbers can also be used to determine the goods that are bought and sold by the organisation. For example, goods that have low sales numbers may be reduced in price and not stocked again, while goods with high sales numbers will be restocked to continue the sale of them to customers.

Where an organisation sells specific goods, these could be classed as a trade secret. Trade secrets often apply to a patent.

> Patents are covered in Content area 4, section 4.1.5.

The IPA also covers software processes in addition to patents for tangible items. This means that if the function of the organisation is to provide cloud-based services then the software processes used by the organisation could be covered by the IPA.

> ### Key terms
>
> *External stakeholders*: groups outside an organisation, for example shareholders.
>
> *Internal stakeholders*: groups within an organisation, for example owners and employees.

> ### Activity
>
> Using the same online retailer as in the previous exercise, define the data that could be held about the goods that are sold. What would the impact on the retailer and stakeholders be if the data was breached?
>
> Create a digital communication detailing your findings. Present your findings to your group.

Pending restructuring

Many organisations carry out restructuring of departments and employees. This may be because of either an increase or a decrease in clients/customers.

However, any leak of the news of a pending restructuring can have an impact on the organisation and its internal and external stakeholders.

If news of a restructure is leaked to employees, it could cause panic. Employees may worry that they could lose their job and may start to look for a different employer. Customers' and clients' confidence in the organisation may decrease and they may look for a different organisation to interact with. This will, obviously, lead to a downturn in finances.

Protecting privacy and confidentiality

One method of maximising the privacy and confidentiality of data is to use access controls, privileges, authorisation and other security procedures to limit the access to the data and information. All important data and information should also be regularly backed up to a secure location to minimise the impact of a data breach. This ensures that any data can be reinstated as soon as possible to keep the organisation functioning smoothly.

> The use of data its access and the impact on organisations and stakeholders is covered in Content area 6, section 6.4.4.
>
> Security processes and procedures are covered in this content area, section 8.2.3.

> ### Test yourself
>
> 1 What is meant by privacy?
> 2 What are the two different types of stakeholders?
> 3 How do clients interact with an organisation?
> 4 Who should access employee salaries?
> 5 Describe one possible impact of a restructure being leaked.

8.1.2 The potential impact of failing to maintain privacy and confidentiality

Every business and organisation stores data and information. What data and information is stored will depend on the function of the business and organisation. The failure to maintain privacy and confidentiality is most often the result of an attack. The attack may take many forms and some of these will be covered in section 8.1.3.

The impact(s) of failing to maintain privacy and confidentiality can be wide ranging but includes financial and reputational impacts.

Financial

The financial impact of failing to maintain privacy and confidentiality can include:

- ▶ possible payment of compensation
- ▶ increased costs to improve security and new computer devices, including installation and maintenance
- ▶ loss of customers leading to loss of revenue
- ▶ loss of revenue if, for example, invoices are lost.

If customer and employee personal data is targeted during the attack, then compensation may have to be paid to the people whose data has been lost. Most of the legislation requires a business or organisation to pay compensation to the people whose data has been lost, stolen or corrupted. This compensation can be very high, sometimes running into thousands of pounds. If more than one person's data has been lost or stolen then this sum of money will, obviously, increase. The cost of compensation may result in the business having to close down if they do not have enough money to either pay the compensation or to carry on trading.

A business may also need to increase the level of security of its digital systems and this can be expensive as new software and hardware may have to be bought. There is the increased cost of the installation and maintenance of this hardware and software.

There is also a high risk that customers will lose confidence and take their custom elsewhere. This can, again, lead to financial worry as without customers, a business may not be able to carry on trading.

The potential loss of data can also have financial consequences. While many businesses and organisations do back up their data, the timing of the backups can lead to data loss. The data that has been lost may not just be, for example, accounts data. It is possible that supplier and customer data has been lost and this may have financial consequences.

There is always a time delay before the most recent backup can be used to reinstate data. This time delay can have a consequence on how the business or organisation is able to interact with customers and suppliers. With many business transactions, for example e-commerce, happening online, customer orders may be lost. The loss of this data can have consequences for customers as their personal data, and the goods they ordered, may be lost.

Emails and contact lists can also be targeted during an attack. The loss of, for example, email addresses and contact information could result in a lack of function, leading to an organisation losing money and possibly going out of business. If customers' data is lost, then this may have a devastating consequence. The loss of personal data could lead to identity theft and a potential financial impact on its victims.

Reputation

If data has been lost during an attack, then the reputation of the organisation will be negatively affected. It is likely that the business will no longer be seen as trustworthy by its customers and their confidence will reduce. This could lead to customers moving their custom to a different business, resulting in the business stopping trading. For example, supermarkets hold details of customers who use their delivery service – if this data was lost, stolen or corrupted, then customers may find another supermarket to do their deliveries.

An attack can result in lost or corrupt data. The business or organisation may have backups, but the time taken to reinstall the data can have an impact on the operations. Data is relied on to carry out day-to-day functions, both internally and during the external interaction between a business and its suppliers and/or customers.

The time delay in restoring the lost or corrupted data will mean that the day-to-day business may not be able to continue, causing disruption to its function. When a business or organisation is recovering from an attack it may not be possible for it to operate because the internal data that needs to be shared between departments, and external data, such as links with suppliers and customers, may be lost or corrupted.

> **Research**
>
> In April 2017, the loan firm Wonga was the victim of an attack. The data of more than 250,000 people who had used the firm was stolen.
>
> Research the impacts on Wonga and its customers of the failure to maintain the privacy and confidentiality of the data.

8.1.3 Potential technical threats and vulnerabilities to systems, data and information

There are many technical threats and vulnerabilities that can have an impact on systems, data and information. It is important that **cyber security** is considered by everyone who uses a digital system. This covers large multinational organisations, governments and individuals. Digital systems store, process and use a wide range of data and information, all of which is important and, if lost or stolen, can have far reaching impacts.

Key term

Cyber security: the practice of defending computers, servers, mobile devices, electronic systems, networks and data from malicious attacks.

Every industry, business, organisation and individual can be the target of technical threats. Every digital system, irrespective of where and why it is used, can also suffer vulnerabilities. The technical threats will vary depending on the nature of the data and information held and the motivation of the attacker.

There are many technical threats and vulnerabilities that can affect systems, data and information. These include:

▶ botnets
▶ DDoS
▶ hacking
▶ malware (including ransomware)
▶ social engineering (pharming, phishing)
▶ insecure APIs
▶ use of ad hoc or open networks
▶ eavesdropping/man-in-the-middle attacks

Bots/botnets, DDoS, hacking, malware and social engineering are covered in Content area 4, section 4.1.3.

APIs

An API is the interface that enables two, or more, different software applications to communicate. Each time a user sends a request or accesses a webpage or app, a remote API is used. Remote APIs can interact through a communications network with the resources, for example a webpage is outside the computer making the request.

The majority of APIs are designed using web standards. Remember: not all remote APIs are web APIs, but all web APIs are remote.

Web standards are covered in Content area 4, section 4.2.4.

APIs are covered in Content area 6, section 6.4.5.

For example, a user is checking the motorway network for delays on their journey. The API enables the user's digital device to send a request to the remote server that stores the webpage. This server then links to the server the data is stored on. When the data is received it is processed by the original server and returned to the user.

Over time, APIs can become unsecure. This can lead to vulnerability that can be exploited by risks and threats. Most data is interconnected as websites, apps and software programs interact. If one API is insecure this can lead to a vulnerability. This vulnerability can lead to a higher risk of threat to everything that this API interacts with.

Ad hoc networks

A network can be created without the use of a wireless router or an access point. This means the digital devices in that network communicate directly with each other – a wireless ad hoc network (WANET).

The main problems with ad hoc networks are a slow data transmission rate and minimal security. If an attacker or hacker is within range of an ad hoc network, they can make a connection to that network very easily.

Research

Research the different types of WANET. Create a digital communication which explains the different types for your peers.

Man-in-the-middle attacks and eavesdropping

A man-in-the-middle (MITM) attack happens when a **hacker** places themselves in the middle of a communication between two digital devices and/or the users. This means that the man-in-the-middle, or hacker, can intercept, send and receive data meant for someone else, or that is not meant to be sent at all, without anyone knowing until it is too late. The hacker can be in the middle of two different users or between a user and a website.

When a hacker is using a MITM attack, they are attempting to steal data and information. This data and information may be log-in details or financial data (e.g. credit or debit card numbers). These details may then be posted to the **Dark Net** where they can be bought and used by other attackers.

Key terms

Hacker: someone who uses computers to gain unauthorised access to data.

Dark Net: networks that are not indexed by search engines. They can only be accessed by people with the relevant credentials and authorisation.

Some hackers may be eavesdropping on conversations carried out over the internet or mobile device communication to gain confidential information which can then be sold for financial gain.

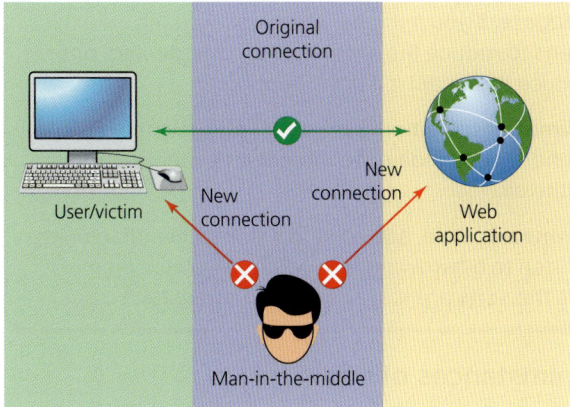

▲ Figure 8.1 A MITM attack

Hacking is covered in Content area 4, section 4.1.3.

Test yourself

1   What does DDoS stand for?
2   What is one problem with a WANET?
3   What is a MITM attack?

8.1.4 Potential physical vulnerabilities to systems, data and information

Physical vulnerabilities can cause threats to digital systems, data and information. The vulnerability will depend on the digital system and the data and information that is stored.

Some organisations run **vulnerability testing**, also known as penetration testing, when the digital system is being created and installed. It is also possible to run vulnerability testing when the computer system is running. These tests can identify vulnerabilities, and steps can be taken to close them before a cyber-security attack is successful.

But the vulnerability that can affect digital devices and systems the most is the users of the computer system. So, users must be aware of the vulnerabilities so that they do not become the start of any issue that compromises the computer devices, system or the data and information.

Penetration testing is covered in section 8.2.3 of this content area.

Physical threats can be split into two categories:
▶   internal
▶   external.

Internal

Location of systems

It is very important to consider the location of digital systems. When selecting the location, the threats need to be, as far as possible, **mitigated** against. For example, the physical locations of the digital systems should be in an area where no flammable material is used or stored, to reduce the possibility of a fire. In addition, an appropriate fire alarm system should be installed. (Many fire alarm systems automatically trigger some level of water to extinguish the fire but this is not a good idea with digital systems that need to be powered by electricity.)

Digital systems generate a lot of heat and, without proper ventilation, this heat can be converted into **humidity**. A high level of humidity can cause internal components of a digital device to corrode or be damaged to the point of not working as intended. It is also possible for components to short-circuit which can lead to a loss of stored data and information.

If the components of a digital system are compromised there is a very high probability that the software the digital system is running will also be compromised. This may lead to authorised users experiencing issues when attempting to access and use the software, and data stores becoming unavailable or slow to access.

Digital systems should be located in a climate controlled area to ensure that the level of humidity can be kept at a low level. They should also be located in areas where there is a limited amount of dust because dust can have the same level of impact on a digital system as humidity.

Layout of systems

The layout of digital systems should also be considered as this can be a vulnerability. For example, individual workstations should have enough of a gap to ensure that cool air can pass between them. It is also important to locate any digital devices away from any windows or doors where they can be seen. The layout of digital systems can be linked with the location of the digital system.

Physical threats can also include:
- door access codes not being updated regularly
- using simple access codes such as 1234
- reusing access codes on a rotation basis
- lack of monitoring of access to secure areas
- unnecessary access to secure areas.

> **Activity**
>
> Carry out an audit of the internal threats in your centre or workplace. Identify any weaknesses and suggest improvements. Create a digital communication detailing your findings.

System or asset robustness

When a digital system is installed, it can be assumed that the hardware and installed software is up to date and can, therefore, be classed as robust. Over the life of a digital system, routine maintenance should be completed on a regular basis. However, eventually software becomes obsolete. This may be because new versions of the software have been released and the patches that enable older versions to run on newer versions are no longer available. It may also be that the software vendor has decided to retire the software.

> **Research**
>
> Adobe issued an end of life (EOL) notice for Adobe Flash Player and stopped supporting it in January 2021.
>
> Research the impact that running an unsupported Flash Player may have on a business's digital system.

This may also cause a problem with legacy software. For example, over the life of a database, the vendor will issue updates to ensure the database software will still perform as required. However, over time the database will need to interact with more up-to-date software. This is when the database becomes a legacy system. At this point a decision has to be made about replacing the database with an up-to-date version which will interact with newer software. However, this decision has an impact on the business. For example, will the stored data be transferred to the new database with a guarantee of no loss of data?

Hardware can also become obsolete and will need replacing. This may be linked with the **firmware** being run on the hardware components or may be as the result of a reduction in performance. It is very common to have updates to firmware issued for devices and components, for example mouses and printers, but there comes a time, as with software, that manufacturers and vendors will stop supporting any given device or component.

> **Key terms**
>
> **Mitigate:** if you mitigate against something, you take steps to reduce the likelihood of it happening, or to reduce its impact if it does happen.
>
> **Humidity:** the amount of water vapour in the air. The higher the humidity, the more water vapour there is in the air.
>
> **Firmware:** code, added at the time of manufacturing, written to a hardware device's non-volatile memory. It is the software that allows the hardware to run.

Circumstances of use

How and why the digital systems, including the data and information, are used could also lead to a threat. Many businesses provide employees with laptops and external storage devices for use both on and off-site. Using these devices and interacting with the digital systems using unsecured hotspots can leave systems vulnerable to attack. There is also a vulnerability related to where the laptops are used and transported.

Guidance about this may be given in an AUP which is read and signed by employees.

User characteristics

Where the digital systems are used by a business it can almost be guaranteed that the employees will have a high sense of responsibility to use them appropriately. However, there is always the chance that a disgruntled employee can pose a threat to the hardware and software of the digital system. But any action can be dealt with under the policies in place, for example, the AUP.

> AUPs are covered in Content area 4, section 4.2.5.

External

With the increase in the use of digital devices and the cloud, there are external, also known as environmental, vulnerabilities that can affect data, information and digital systems.

If a natural disaster occurred, for example an earthquake, then it is probable that internet access would be lost. This could mean that any data and information stored on the cloud would be inaccessible. The impact of inaccessible data and information could also affect the recovery from a natural disaster.

It would also be possible that digital devices could be destroyed during a natural disaster. If a tsunami or flood happened, the water coming onto the land could destroy or wash away buildings. If digital devices were in these buildings, then they would be destroyed or lost. The cabling infrastructure or any internet service equipment could also be affected. Even if buildings could be made safe, the tremors that can happen with a natural disaster, such as an earthquake, could damage any hard drive surfaces causing the data and information stored on them to be unreadable.

Even if physical backups were available, there is a probability that these would also be affected by the same natural disaster. If the backups were stored on the cloud, then these may also be inaccessible as there may be no internet access.

Power failure is one of the potential after-effects of a natural disaster. As digital systems need electricity to either charge or operate, this will also mean very limited access to data and information, and the digital devices these are stored on. One method that can be used to keep digital systems operating could be to use batteries or a power generator as backup power sources. But the batteries must be kept fully charged and fuel must be available to run the generator.

Lightning strikes are another natural disaster that can affect computer systems and devices. A lightning strike can cause a surge or spike in the electricity supply. These surges can affect how the hard drives and other storage devices operate.

To mitigate against physical threats, physical security measures should be implemented.

> Processes and procedures that can be used to mitigate against threats are covered in section 8.2.3 of this content area.

Research

Investigate the different devices that can be used to protect against power surges. Identify where each device could be used. Make notes about your findings.

Test yourself

1 What is vulnerability testing?
2 Why should digital systems be located in a low humidity location?
3 What is a legacy system?
4 What is firmware?
5 Identify two types of environmental threat.

8.1.5 Potential human threats and vulnerabilities to systems, data and information

You have already learned that people are one of the biggest threats to digital systems, data and information. Human error was the cause of approximately 90% of data breaches in 2019 according to statistics from the ICO (Cybsafe, 2020).

Research

In 2008 Facebook made public the dates of birth of 80 million users. How did this happen? What are the possible impacts of this breach on Facebook users? Make notes about your findings.

Human threats include:

▶ human error
▶ malicious employees
▶ disguised criminals
▶ targeted attack.

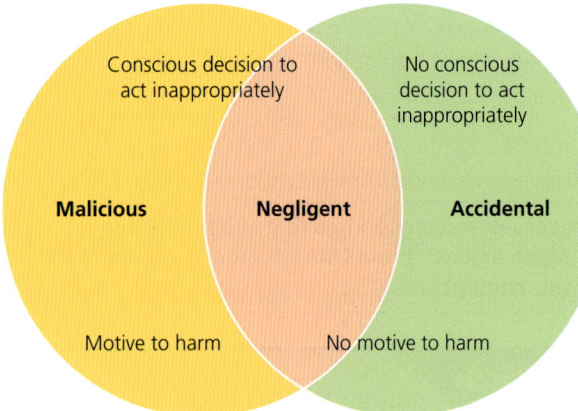

▲ Figure 8.2 Types of human threats

Figure 8.2 shows the differences between malicious, negligent and accidental human threats to digital systems, data and information.

Human error

Human error can lead to an accidental loss of data. This is a loss of the data itself rather than a loss of a copy or backup version of the data. For example, the loss of a hard copy of the data would not result in the loss of the source of that data.

Human error can include:

▶ accidentally deleting a file containing the data, or shredding the final hard copy of a data file
▶ saving files and folders to a different location
▶ sending emails to the wrong recipients with attachments containing data
▶ accidentally making changes in documents.

While every person is capable of making an error, businesses and organisations should attempt to minimise the likelihood of these errors happening. This may be through the use of regular employee training, high-profile reminders to employees, for example on splash screens on digital devices, and ensuring that all policies and procedures are read and understood by all employees.

Malicious employees

Malicious employees can be another threat to digital systems, data and information. Malicious employees

are also known as Turncloaks. They typically use their access details in a malicious and deliberate way to steal information and data for financial or personal reasons. An individual may become a Turncloak as a result of a social engineering attack.

While many employees take no further action if they are disciplined or sacked, a Turncloak employee will hold a grudge against their employer. This type of threat is often difficult to trace as they are familiar with the security procedures of the business as well as any vulnerabilities.

> **Research**
>
> In 2015 a US health insurance company, Anthem, suffered a data breach. Social engineering was thought to have provided the access codes to the customer database.
>
> Identify the different types of social engineering and describe how each type could have been used to gather the required access codes. Make notes about your findings.

> Disguised criminals (social engineering) are covered in Content area 4, section 4.1.3.
>
> Targeted attacks (hackers) are covered in Content area 4, section 4.1.3.

8.2 Threat mitigation

8.2.1 Understand the concept of the CIA and how it can be applied to define security aims

Security, in particular cyber security, aims to protect digital systems, data and information. Cyber security attempts to:

▶ act as a deterrent against attackers and hackers
▶ prevent an attack from happening
▶ detect and warn users of the digital systems that an attack is happening.

The main purpose of cyber security is to maintain the **confidentiality, integrity and availability (CIA)** of digital systems, data, and information.

> **Key term**
>
> ***Confidentiality, integrity and availability (CIA):*** also known as the CIA triad.

▲ Figure 8.3 The CIA triad

Figure 8.3 shows the CIA triad, viewed here as a triangle with security in the centre. The CIA triad is a security model developed to define the important parts of cyber security and how they are interlinked.

- **Confidentiality** means that the digital systems, data and information resources are protected from unauthorised viewing and access (**hacking**).
- **Integrity** means that data is protected from unauthorised changes to ensure that it is reliable and correct.
- **Availability** means that authorised users have access to the digital systems, data and information they require.

The CIA triad shows the clear relationship between these three parts of cyber security. Looking at these in a triangle we can see that they overlap, but they can also work against each other when deciding which types of mitigation to use. Visualising things in this way enables an organisation to plan and prioritise the implementation of new security policies and processes.

A good example of how confidentiality, integrity and availability interact can be found in online banking:

- Confidentiality – it is important to a customer that their financial details are kept confidential between them and the bank. One strategy that can be used to maintain the confidentiality of the financial data is through access-level login. When a customer logs into the bank website their login details provide access only to their bank account (and no one else's).
- Integrity – the financial data of the customer must demonstrate integrity. This means that the customer can expect their financial data to be correct. For example, their recent transactions

using their debit or credit cards should be true and accurate. The financial data should also be reliable, which is linked to its accuracy.
- Availability – customers should be able to access both the bank website and their financial records when they want and need to. If the website or personal financial data is not available, then this part of the CIA triad has been broken.

8.2.2 The interrelationship between security, identity, confidentiality, integrity, availability, threat, vulnerability and risk management within a business context

Security aims to protect digital systems, data and information. Part of this is to ensure that the digital systems, data and information are not compromised when/if a critical threat happens.

By using security, the likelihood of a threat being successful is reduced because the identified vulnerabilities of the digital system, data, information and people will also be reduced.

Security must be used to maintain the CIA triad. There is a strong relationship between all the different components, but using security reduces the chance of any of the components being compromised.

8.2.3 Processes and procedures to mitigate threats and ensure security

Data and information are very valuable assets, not only to the businesses and organisations that collect, store, process and use them, but also to each individual.

Data and information, such as customer shopping records, financial data and health data and information, are used for a variety of purposes. What is important is that all data and information are kept secure and protected from the large range of threats that could occur.

Some of the ways that threats can be mitigated against include:

▶ air gapping
▶ anti-virus and anti-malware programs
▶ certification of APIs
▶ configuration and management of software-based access control
▶ device hardening
▶ encryption
▶ user access restrictions
▶ multi-factor authentication
▶ firewalls
▶ password managers
▶ policy, policy enforcement and training
▶ SYN cookies
▶ virtual private networks (VPNs)
▶ security testing (penetration testing, white/grey hat hackers).

Air gapping

Air gapping is a digital system that is physically isolated from potentially dangerous networks, such as the internet. Basically, air gapping is having a digital system that works offline.

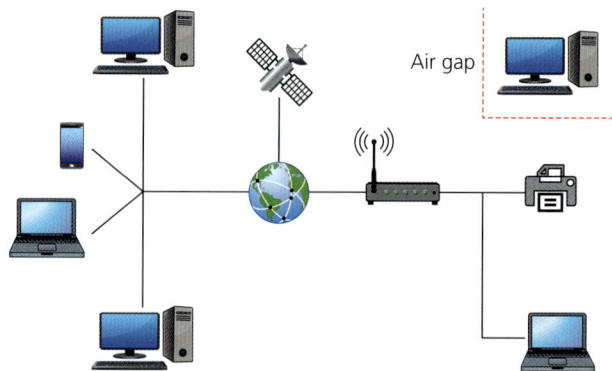

▲ Figure 8.4 An air-gapped digital system

As shown in Figure 8.4, an air-gapped digital system is one that is not connected, either physically or wirelessly, to other systems or networks. It is usually a standalone system or a network of digital systems that has no external links to any other system.

Air gapping refers to the concept that there is air between the digital system and any other system or network, including the internet. This means that the air-gapped system cannot be the victim of a threat or attack through another network. To carry out an attack on an air-gapped system would require the attacker to be physically sitting at the system.

There are still threats to an air-gapped system. The main threat is the use of removable storage devices. For example, a user downloads an infected file from a network onto a USB memory stick. The memory stick is then used to upload the infected file to the air-gapped system. This means that the air-gapped system is now infected and has been the victim of a threat.

However, to some businesses and organisations, using the air-gapping technique to mitigate against threats is not always feasible. The reason digital systems are used in business is because they can share information and data, and access this data and information, from a centralised storage area.

But air gapping, if done properly, can provide complete protection to the air-gapped digital system The other main advantage to using an air-gapped digital system to mitigate against threats is that once the air gapping has been carried out, there are no ongoing, recurring costs.

Anti-virus and anti-malware

Anti-virus and anti-malware programs are security software which are designed to prevent, detect and remove viruses and other malware, including adware, Trojans and worms. It is essential that any digital system connected to the internet has some form of security protection. If security software is not installed then it is possible that within minutes of connection to the internet it will be infected.

Security software scans files and directories for viruses or malware. If malicious code is detected then the software will delete it – either after seeking permission to do so from the user or automatically. Security software can also identify, and warn users about, unsafe websites or suspicious emails.

When security software finds a malicious program on a digital system, the user is usually offered two options:

► to quarantine it so the software cannot infect the digital system – this option gives the vendor the opportunity to analyse the program so that they can offer an update to users
► to delete it – this option clears the digital system of the infection.

Automatic versus manual updates

Some security software updates automatically. This process is usually completed in real time. This means that when the computer system is connected to the internet the security software will automatically be checking all the time for new updates. If an update is found, then the security software will automatically update it. This happens because new viruses, and other security threats covered by the software, are being released all the time.

This means that the user does not have to remember to manually check for updates and so the digital system is always protected from any threats.

If a business uses automatic updates of security software then they do not have to remember to manually check for updates and can be sure that their digital system is as up to date as possible. This also means that any vulnerabilities identified by the vendor are solved before an attack can take place.

Manually updating security software can be dangerous to the digital system and the data and information

held on it. Employees can forget to carry out manual updates and this can leave the digital system vulnerable to threats.

A manual update for security software could be completed on an ad hoc basis or can be set to check at a specified time by a user.

One of the problems with manual updating of security software is the time it can take to download the patch. A second problem is that there may be a time delay between the patch being released by the software vendor and the time when the manual update takes place.

Another problem with manually scheduling an update is that the digital system must be switched on and connected to the internet for the update to be downloaded. If the manual update has been scheduled for a time when the business system is switched off, then the business will never get updates or download patches. This can leave the digital system open to attacks and threats and could result in data being lost or stolen.

Some users, however, may prefer to update their software manually because they want to look at the updates to decide whether or not to download them. Some users may consider the updates to be intrusive or not appropriate.

Research

Choose any two of the different providers of anti-virus software – look on the internet to see the different providers available.

Copy and complete this table to show the features which are available (two features have been given for you). You may need to add more rows to the table.

	Provider 1	Provider 2
Internet links scanner		
Live support		

Certification of APIs

Many businesses and organisations use an API. An API enables two, or more, software applications to interact. APIs enable a business or organisation to allow access to their assets, including data and information, while still maintaining the security of their assets. To maintain the security of assets the API has to be regularly maintained and managed. One way that an API can be used to maintain security and mitigate against threats is to set a certificate.

APIs can fall into three different levels of certification – private, partner, public.

▶ The **private** certification means that the API is only used internally within the business or organisation. This is the most secure certification as the assets can only be accessed within the internal network.
▶ The **partner** certification means that the API is only available to trusted partners of the business or organisation. This certification is secure as long as the partners can be trusted and have a high level of security on their own systems.
▶ The **public** certification is the least secure. It means that the API is available to everyone. For example, third parties can develop apps that can interact with the API. This means that security risks on the app could, through the interaction, have access to the assets.

Configuration and management of software-based access control

Software-based access controls aim to mitigate against threats by predefining access by authorised employees, that is which areas of the building each employee can and cannot go into. Most workplaces will need to implement a networked access control system.

This type of system is used where control is required at a central point, for example the reception area. The system will enable access control for employees through a number of doors. Each employee can be provided with a specific access control so that they are able to gain access to the areas that are needed for them to perform their job functions. This control will restrict access to other areas that they do not need access to. Having an access control system can go some way to avoiding the threat of unauthorised access which could lead to theft, malicious damage and threats to personal safety.

The most up-to-date control systems will enable reports to be run to, for example, identify which employees have used a door. These systems can also be integrated with other security controls such as CCTV and alarms.

Employees can gain access through doors by inputting a unique token which can be:
▶ a numeric code/PIN input using a keypad
▶ an access badge using **radio frequency identification (RFID)**
▶ possession-based authentication.

Door access control instructions are completed centrally using a digital device and are then sent to each of the doors.

A unique token can be stopped from having access to all doors instantly. This could be very helpful if an employee is sacked. A NAC system should enable different access permissions for employees and, possibly, at different times of the day.

For example, only those employees who need access to the HR department will be granted the unique token to access this department during working hours. Night security employees might be able to access all doors between specified areas.

Flexible control allows for different access permissions to be granted for individuals or groups of users and at specified times of the day.

As stated, the access control system will log which employees have accessed which doors. This could be very valuable in the case of a fire leading to an evacuation of the workplace. The log will enable employers to find out quickly if any employees have not left the building. This information can be given to the Fire Service who can then enter the building to find the missing employee(s) and save lives.

> Possession-based authentication is covered later in this section.

Key term

Radio frequency identification (RFID): tiny chips that contain information which is transmitted when near a receiver.

Test yourself

1 How does a software-based access system protect a workplace?
2 Identify the three different types of tokens that can be used in a software-based access system.
3 What does RFID stand for?
4 How can flexible control be used?
5 How can an access control log be used?

Device hardening

Device hardening aims to reduce the vulnerabilities of a digital device. By reducing those vulnerabilities, the risk of a threat or attack happening is also reduced. The main aim of device hardening is to reduce as many risks and threats to a digital system as possible. The ideal time to carry out the process of device hardening is before the device is installed onto a network. By doing this, the device will have no potential vulnerabilities that can be exploited by attackers and will not compromise the network.

The process includes activities such as:
- turning off any non-essential features
- updating and installing security patches/updates
- configuring security controls such as password management and file permissions
- disabling unused network and external device ports.

Research

Research other device-hardening activities. Try to find at least three additional examples.

Research

Research the benefits of using device hardening. Create an infographic to present your findings to a group of people who have little knowledge of IT security.

Test yourself

1 What is the aim of device hardening?
2 When is the ideal time to carry out device hardening?
3 Identify three activities associated with device hardening.

Encryption

As has already been discussed, the most valuable assets to a business, organisation or individuals are data and information.

Encryption can help to prevent the data being accessed and used by unauthorised people (attackers). Data can be scrambled by using encryption software when it is stored or transmitted between digital devices over networks.

Data encryption software uses an **encryption code or key** to scramble (encrypt) the contents of data files. The proper code is needed to unscramble the file (decrypt it) so it can be read and used. If the encrypted file is accessed by anyone without the proper code to unscramble it, the data is meaningless.

Key terms

Data encryption software: software that is used to encrypt a file or data.

Encryption code/key: a set of characters, a phrase or numbers that are used when encrypting or decrypting data or a file.

Data at rest: data stored on a digital device or storage medium.

Data in transit: data being sent to one or more authorised users.

Data can be encrypted at rest and/or in transit. It is common practice, especially where the data is sensitive, for example financial or personal data, to encrypt the **data at rest**. This can help to mitigate the loss of the data against an attack.

It is also good practice to encrypt the **data in transit**. By doing this the encryption can mitigate against the loss of the data while it is being transmitted by, for example, email or uploading to a cloud storage area.

Asymmetric and symmetric encryption

There are two main types of encryption: asymmetric and symmetric.

Asymmetric encryption is also known as public key encryption. This is when the encryption key is available to anyone to use and encrypt data but only the person who receives the data receives the decryption key.

Figure 8.5 shows the process of **symmetric encryption**. This is when the encryption and decryption keys are the same.

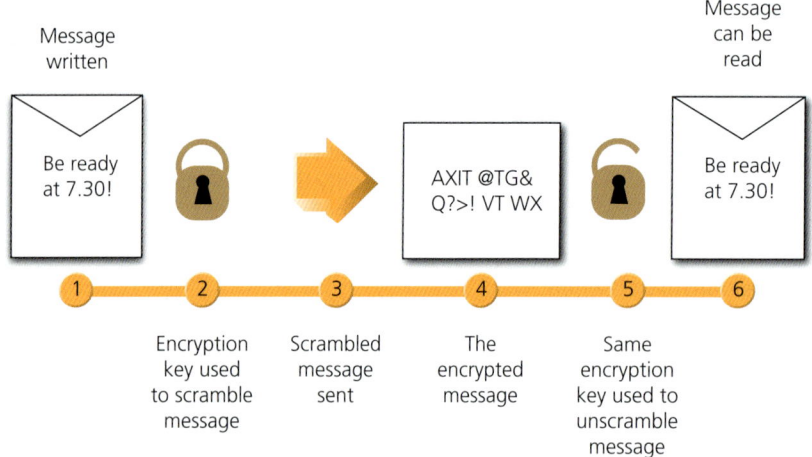

▲ Figure 8.5 Using an encryption key to encrypt and decrypt a message

Encryption can also be used on websites. When customers buy goods online, book cinema tickets online or enter personal details into any website, the data should be encrypted before being transmitted. This will keep the details from being read or used by others even if they are intercepted. Everyone should check that the website they are using to enter personal details uses encryption.

A secure website using encryption will use https instead of http in the URL and will show a small padlock. Different web browsers will show the use of https in different ways.

Figure 8.6 shows the web address beginning with https and also shows the padlock symbol to confirm that the website uses encryption.

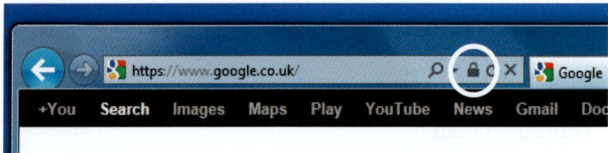

▲ Figure 8.6 A website using encryption

Another method of encrypting data is to use hashing. There is a difference between encryption and hashing.

Encryption is, as already discussed, a two-way process: what is encrypted can be decrypted with the proper key. So, this means that encryption is reversible.

Hashing

Hashing uses an algorithm to map, or scramble, data of any size to a fixed length. This is called a **hash** value. Hashing verifies that a file or piece of data is authentic and has not been altered.

Hashing can be particularly useful for storing passwords. Hashing can also be used for searching, for example to find specific data in a very large database, or for cryptographic applications in digital certificates.

Unlike encryption, hashing is only one way and is not reversible. However, technically it is possible to reverse hash, but this would take vast amounts of processing power and so it is generally considered to be infeasible.

User access restrictions

User access restrictions can be used to mitigate against threats by limiting access to data and information, and physical rooms, based on job roles. There are many different types of user access restrictions that can be used. These include:

▶ logical
  - usernames
  - passwords and passphrases
  - data access levels and permissions
▶ physical
  - physical access control and restrictions.

Usernames

Usernames are part of the log-in credentials provided to employees by their employers. The username, when linked with a correct password or passphrase, is used to provide access to digital systems, data and information. Usernames can also be known as a log-in ID or user ID. They are unique within a workplace. If two users had the same username then this could cause issues when setting access rights and permissions.

Usernames allow lots of users to use the same digital system. The username will enable a user's personal settings and files to be shown.

Passwords and passphrases

Passwords and passphrases are linked with usernames and complete the log-in credentials. The log-in credentials provide protection on two levels. In a workplace the username allows access/gives permission for the user to access specific software such as financial or HR.

The password, or passphrase, can allow the user **access rights** and **permissions** to the digital system and software such as internet and email access, and standard office applications.

If the data is extremely sensitive then further passwords, or passphrases, may be needed.

Passwords

It is also good practice for external storage devices, files and folders to be password protected. This is only effective if the password is strong and not easily guessed.

Passphrases

Passphrases are very similar to passwords but are usually a string of words – a phrase. There are some advantages to using a passphrase rather than a password.

Passphrases are:

▶ usually longer than a password, meaning they are more difficult to be guessed by an attacker (most password-cracking software used by attackers has a limit on the number of characters they can be used on)
▶ random words rather than a standard well-known phrase
▶ more memorable to the user because a passphrase can be easily remembered, so limiting the number of 'forgotten passphrase' requests by users.

As with passwords, it is good practice not to use the same passphrase over multiple log-ins.

Data access levels/permissions

Log-in credentials can be used, as already discussed, to provide access levels and permissions. Digital systems, software, files and folders can have access rights and permissions set. This means that only those users who have the correct log-in credentials can have access.

The access rights can inform what a user can do with a file and folder. These are called permissions. A user can have permission to read, write, edit or delete data and information.

> Access rights and permission are also covered in Content area 6, section 6.4.4.

Physical access control/restrictions

There are many different types of physical access control/restrictions. The ones that are implemented will depend on the function of the business or organisation and the physical workplaces.

There are many different types of access controls that can be implemented including:

- access badges
- alarm systems
- external security lighting
- barriers
- CCTV
- door sensors
- locks and keypads.

The physical security measures used will depend on the physical layout of the workplace. If attackers gain access to a workplace then they could steal physical digital devices and could also infect digital systems with malware. How successful this would be will depend on the logical security measures that have been implemented.

> Some details about access badges are covered in Content area 4, section 4.1.6.

Alarm systems

These can be installed to alert people to an unlawful attempt to access the workplace. Different types of alarm systems can be used including motion sensors and glass break detection. Alarms can also be silent or have an audio signal that the alarm has been tripped.

However, an alarm is only fully effective if there is a quick response to the alarm as this will increase the chance of catching the attackers.

Barriers

Barriers can be used to access any car parks or distribution areas in a workplace. They can limit access until a code or verbal authorisation has been given. If authorisation is provided, then the barrier is lifted and access is granted.

External security lights

These are a relatively inexpensive and quick method of increasing security effectively. Lighting is often a deterrent for attackers as the mere threat of potentially being seen is more than enough to stop

them attempting to gain access to a workplace. If the external security lights are automatic/motion sensing, they will turn on automatically if any motion within the range of the lights is detected.

CCTV

CCTV can be a very strong deterrent against any unlawful activity. As with external security lighting, they are quick to install and can be inexpensive. CCTV cameras can provide video footage of criminal activity but, as with security lights, just the sight of them can be a deterrent to attackers. Signs stating that CCTV is being used must also be installed.

> **Research**
>
> Research the physical protection methods of access badges, door sensors, and locks and keypads. Make notes about your findings.

> **Activity**
>
> Carry out a physical security analysis of your centre or workplace. Create a digital communication detailing any potentially insecure areas and recommending how the physical security could be improved. The digital communication should be aimed at a senior management team (SMT).

> **Test yourself**
>
> 1 Identify two types of physical security methods.
> 2 What is the main disadvantage of alarm systems?
> 3 How can barriers limit access to a workplace?
> 4 What has to be in place if CCTV is installed?
> 5 How do automatic security lights work?

Multi-factor authentication

Multi-factor authentication can take many forms, including:

- possession-based
- knowledge
- biometric
- location-based.

Multi-factor authentication is also known as two-factor authentication (2FA) or three-factor authentication (3FA).

Possession-based

Possession-based multi-factor authentication can also be known as token authentication. This method is based on a possession that a user typically has with them at all times, for example a mobile phone.

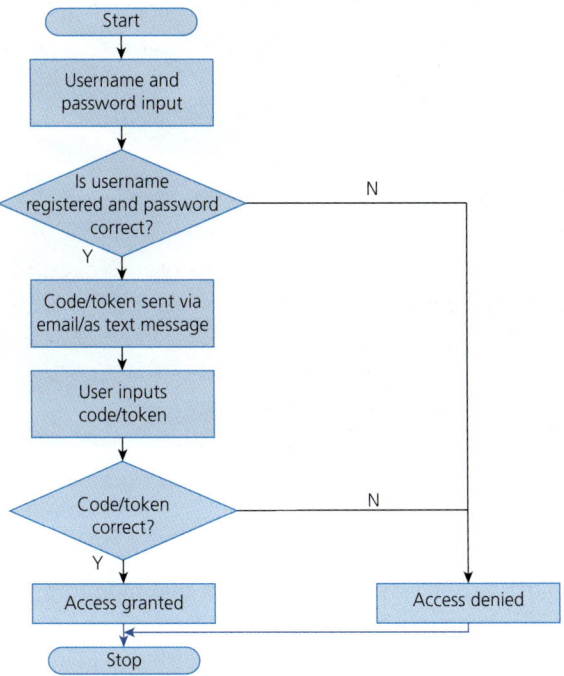

▲ Figure 8.7 Two-step authentication

Figure 8.7 shows the process used during possession-based multi-factor authentication. When a user needs to enter a secure area of a digital system, the log-in credentials are input. When the log-in credentials have been submitted, they are checked by the digital system. A token code, usually numeric, is sent to either by email or text to the email and phone details linked to the log-in credentials. The email address and mobile phone number will be stored on the digital system. The user receives the token code and, to access the secure area, inputs this code.

Log-in credentials are covered earlier in this section.

Biometric

Biometric multi-factor authentication uses a person's physical characteristics, for example a fingerprint, eye scan or voice.

It is common for laptops, smartphones and tablets to need a biometric measure to be positive before these devices can be accessed. The owner of these devices will have stored their characteristic as part of the security settings on these devices. When, for example, the fingerprint is used to access the device, this is checked against the stored fingerprint characteristics and if there is a match then access is granted. This means that only people whose characteristic is stored and recognised can access the device. If anyone else tries to access the device, the characteristic will not be recognised and so access will be denied.

Large businesses can use biometric protection measures to protect, for example, server rooms. When someone tries to access this room, they will scan their characteristic, for example their fingerprint. This is then checked against the database of authorised personnel fingerprints and, if there is a match, then access will be granted. This means that access to areas or rooms of a workplace can be limited.

There are some disadvantages to using biometric protection measures. For example, a person's voice can change if they have a cold. This can cause problems if the device they are trying to access does not recognise the voice pattern. People can have an injury to their fingers, for example, a burn or a cut. If the injury is severe, this can change the pattern of the fingerprint and may result in access being denied. Another example may be if someone has been swimming and their fingers get wet and wrinkly. This will change the pattern of the fingerprint and will result in the scanner not recognising the fingerprint.

Knowledge

Knowledge-based authentication (KBA) authenticates a user by using questions and answers which have already been agreed. KBA relies on the user having to prove their identity by sharing information about themselves through answers to questions. The most commonly used KBA in the UK is static.

Static KBA is commonly known as secret questions. If, for example, a user forgets their password or passphrase of their log-in credentials, then the secret question can be used to verify the identity of the user. Answers to secret questions are usually provided when an account is set up – often the user can select the security question from a list of options. The answer to the security question is linked to the log-in credentials and stored in a file with a high level of security.

Research

Create a list of secret questions that could be offered to users of a cloud storage area. Discuss your questions with the rest of your group.

Location-based

Location-based multi-factor authentication authenticates a user through their physical location. If a user wants to log into a digital device which is wired into a network, then a specific PIN could be used.

This is because, as the device is wired into a network, the location of the user can be verified. Another factor could be that access to the workplace is controlled by physical security such as access badges.

If, however, a user wants to use the network remotely then log-in credentials and a token code would be required.

Another method of location multi-factor authentication is that of verifying a user's location via IP addresses. For example, many web-based services use geolocation security checks.

When an account is set up, an address is usually required. The address will include the county and country. If an attacker attempts to log into the account from a different location, then the registered account holder will be notified.

▲ Figure 8.8 A firewall acts as a barrier against threats to a system's security

There are two formats of firewall:
▶ A **software firewall** is a program that monitors traffic through port numbers and applications.
▶ A **physical firewall** is a piece of hardware installed between the network and the gateway.

It is advisable to use both software and hardware firewalls in tandem, in order to increase their efficiency. Both monitor incoming traffic and analyse it against set security rules. Any traffic that breaks those rules is blocked.

Firewalls monitor the traffic at the entry point – called ports. This is because ports are where the information is exchanged with the external devices.

There are three main types of firewalls:
▶ **Packet filtering** firewalls mitigate against threats by analysing the data packets and blocking any packets that do not meet the predefined security rules.
▶ **Proxy** firewalls mitigate against threats by taking on the role of the intended recipient. They monitor traffic at the application level. Proxy firewalls monitor traffic for seven-layer protocols, including HTTP and FTP.
▶ **Inspection** firewalls mitigate against threats by marking the key features of any outgoing requests for information, checking for the same key features of the data coming into the system and deciding if the incoming traffic is relevant.

The seven-layer protocols are covered in Content area 7, section 7.2.9.

Test yourself

1 What does the abbreviation 2FA mean?
2 How can a token code be sent using possession-based authentication?
3 What would be required if a user was logging into a network using a wired connection?
4 Identify two features that could be used for biometric authentication.
5 What is meant by static KBA?

Firewalls

A firewall is a security device that mitigates against threats by examining **data packets**. A firewall can be either hardware or software or both, but hardware and software firewalls work in the same way.

The purpose of a firewall is to establish a barrier between a digital device and/or a network and incoming traffic from external sources (such as the internet). Firewalls monitor the traffic that flows into a digital device and/or a network through an internet connection. The firewall blocks malicious traffic like viruses and hackers based on security rules.

Key term

Data packets: small units of data which are sent and received when accessing the internet or any other type of network.

Research

Research NGFW, NAT and SMLI firewalls. Create an infographic aimed at 17 to 18 years olds to explain how each of the firewalls works.

Password managers

A password manager mitigates against threats by generating, retrieving and keeping track of randomly generated passwords across various account log-ins and online services. A user can use this to store their existing passwords, in an encrypted format, or to generate new ones.

Most people, in both their working and personal lives, use weak passwords or reuse passwords on multiple accounts. By doing this there is an increased risk of threats, including identity theft. A password manager may also store PINs and credit/debit card numbers with the three-digit CVC code (found on the back of credit and debit cards). Answers to security questions may also be stored by some password managers.

One advantage of using a password manager is that only one password needs to be remembered – the password to the password manager. Another advantage is that if a cloud-based password manager is used, the manager can be accessed from any digital device with a connection. Another advantage is that using a password manager can save time. Some password managers can auto-fill details, including name, address, email, phone number and payment details for faster access to online accounts.

Password managers can also auto-generate secure passwords. Users may have the option to use an auto-generated password when they are creating a new account for a website or application. The auto-generated passwords are usually long, and include alphanumeric characters meaning they are difficult to guess.

If the password manager is cloud-based then an alert can be shown if a website is fake, usually as a result of a phishing attack. Phishing emails characteristically contain a link to a fake website which looks very similar to the real website. A web-based password manager will not auto-fill log-in details as it does not recognise the website as the website the log-in details are linked to.

> Social engineering, including phishing, is covered in Content area 4, section 4.1.3.

Policy, policy enforcement and training

One of the most important processes and procedures to mitigate threats and ensure security is to ensure employees take all forms of security seriously.

An AUP should include the processes and actions that employees must follow and take in the event of any security breach.

> AUPs are covered in Content area 4, section 4.2.5.

Policy enforcement

It is important to enforce any policy created and used by a business or organisation. All employees must read and understand the policies that are in place and an incident response (IR) policy should also be created. Part of this policy should cover the actions to be taken if a security attack has happened. Employees should be encouraged to report to management any activity that raises suspicions.

If a security attack has occurred, then an incident report should be completed. This report could set a precedent and provide details about how to respond if a similar attack happens in the future.

Many businesses and organisations will already have an incident report policy which will provide information including details of the key employees and decisions.

If an attack has occurred and the incident report needs to be completed, the key elements of the report should include:

▶ The title, date and time of the incident – these are important as they can be cross-referred to, for example when updating and installing any vendor security patches, or anti-virus and anti-malware software.

▶ The target of the attack – did the attack just target a specific department, for example finance, or was the whole business or organisation targeted?

▶ Incident category – what was the severity of the attack? An attack can normally be categorised as critical, severe, significant or negligible.

▶ A description of the incident – this should cover exactly what the problem was, how it was identified and details about the attack.

▶ Type(s) of attacker – did the attack originate externally, internally, a single attacker or a group?

▶ Purpose of the incident – what was the aim of/ motivation for the attack?

▶ Techniques used – what was used – for example social engineering, DDoS, unauthorised access to a workplace?

▶ Impact of the incident – for example, was sensitive data accessed, is the business reputation intact or has it diminished?

▶ Cost of the incident – what did it cost to recover from the attack, in terms of finance, working time, reputation?

▶ Response needed – was or does the incident need to be reported to the relevant authorities, for example the Police, ICO, software vendors.

▶ Future management – a review of the incident and an evaluation of the findings to, for example, identify any trends, review current policies and procedures, any recommendations.

Training

Employees should attend regular training sessions or courses. These are particularly important if data and information is stored and/or much of the operational procedures are completed online. Security threats and the associated legislation continually evolve so it is important that all employees keep as up to date as possible.

Training undertaken by employees could include information about the latest security threats, basic security knowledge such as how to spot fake websites, and phishing and pharming attacks. Employees should also be aware of the basic security processes such as not downloading attachments from unknown source emails.

Phishing and pharming are covered in Content area 4, section 4.1.3.

The different areas of security knowledge required by employees depend on job roles. However, it is important that all employees are encouraged to be vigilant and report anything unusual, for example, visitors who do not have the appropriate badge.

Test yourself

1 Identify two sections of an incident response report.
2 What is meant by the significant incident category?
3 Why should an incident report be completed after an attack?
4 Identify one basic security process.
5 Why should employees undertake regular training?

SYN cookies

SYN cookies mitigate against threats by attempting to combat **IP spoofing**.

They work by attempting to stop an attack on the three-way handshake with the TCP server which handles requests. Figure 8.9 shows the process of the three-way handshake.

Key term

IP spoofing: changing a packet's source IP address to impersonate another computer system, or to hide the identity of the sender, or both.

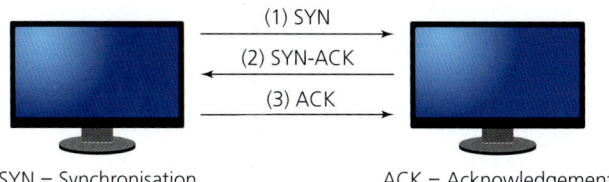

SYN = Synchronisation   ACK = Acknowledgement

▲ **Figure 8.9 The three-way handshake with a TCP server**

A **SYN packet of data** is sent to the TCP server. The server **ACK**nowledges the SYN packet and sends back a SYN-ACK packet of data. A final ACK packet is then sent which enables the request to be processed.

Attackers use this process by sending **false SYN packets** to the targeted server. The server will then respond with a SYN-ACK packet to these false SYN packets, keeping a port open ready to make a connection, and wait to receive the final ACK packet.

However, the final ACK packet is never received by the targeted server but there is an open connection ready which can then be exploited by the attacker. The most common type of attack carried out using this method is a type of DDoS called a SYN flood.

SYN cookies can be used to avoid attempted DDoS attacks. A SYN cookie is created by a server. This enables the sever to create a SYN-ACK packet but the SYN request is dropped and removed from the server memory. The port is open and ready to make a connection.

If a final ACK packet is received from the source, meaning the request is from an authorised source, then the SYN-ACK request is reconstructed. However, through this reconstruction some information may be lost relating to the TCP connection request. This is, however, better than being the target of a DDoS attack with the server becoming flooded with false SYN requests making it unusable for authorised users.

The TCP model is covered in Content area 7, section 7.2.10.

Test yourself

1 What do SYN cookies aim to do?
2 What is the three-way handshake?
3 What happens when the final ACK packet is received?
4 What is the result of the final reconstruction of the SYN-ACK?
5 What happens when a DDoS is carried out?

VPNs

VPNs mitigate against threats by creating a secure connection to another network over the internet. This makes it possible to establish a private, safer and more secure network from a public internet connection.

VPNs were created as a method of connecting business networks together securely over the internet or to allow employees to access a business network remotely. VPNs also mask the **Internet Protocol (IP)** address, which means that online actions are virtually untraceable.

When a digital device is connected to a VPN the device appears to be in the same local network as the VPN. For example, if the VPN is based in Australia then it will appear that the connection is coming from Australia.

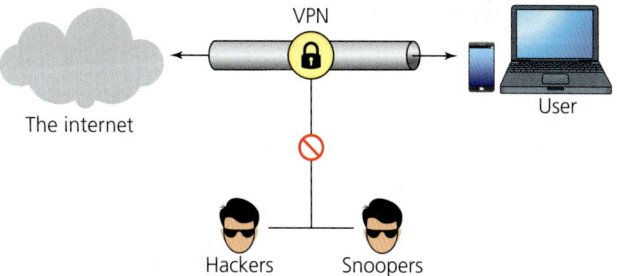

▲ **Figure 8.10 A VPN works to protect a user from threats by creating a secure connection to another network over the internet**

Figure 8.10 shows how a VPN connection works. All traffic is sent from the digital device over a secure connection to the VPN private server. So, when the internet is browsed on the digital device, the website requested is contacted though the VPN. The VPN hides the device IP address, protecting the identity of both the device and the user. The VPN forwards the request and sends the response from the website back through the VPN secure connection to the device. A VPN creates a private tunnel from a device to the internet and hides data through encryption. It is also possible to set access controls on a VPN if it is being used by a business or organisation. By doing this, access to data stored on, for example, the cloud, can be limited to those who need access for their job role.

As can be seen in Figure 8.10 using a VPN makes it more difficult for attackers to access the information and data being transmitted. This is because if the data is intercepted then it will be unreadable until the final destination is reached.

Research

Investigate available VPN providers. Produce a presentation, including speaker notes, comparing the features they offer. Your presentation should help a business to compare different providers before selecting which one to use.

Test yourself

1 How does a VPN mitigate against threats?
2 What do VPNs aim to mask?
3 Define IP.
4 Where does a device appear to be located when connected to a VPN?
5 Why is it difficult for attackers to access data that is being transmitted?

Security testing

All businesses and organisations that use digital systems should carry out security testing on a regular basis. By doing this, they can identify vulnerabilities and rectify them before an attacker exploits them.

Penetration testing (also known as **ethical hacking**) can be carried out by white or grey hat hackers.

Grey and white hat hackers are covered in Content area 4, section 4.1.3.

The **NCSC** defines penetration testing as:

> A method for gaining assurance in the security of an IT system by attempting to breach some or all of that system's security, using the same tools and techniques as an adversary might.

There are many different types of penetration testing that can be carried out.

Key terms

Internet Protocol (IP): the string of numbers an ISP assigns a device.

Ethical hacking: an alternative term for penetration testing.

NCSC: The UK's National Cyber Security Centre.

Network penetration testing

This can be carried out to look for internet and/or external openings to identify how the vulnerabilities could be exploited by internal and/or external attackers.

A network attack is the most common type of penetration test. Network attack tests may include analysing network traffic, testing routers, and identifying legacy devices and third-party appliances where updates have not been implemented.

Social engineering penetration testing

This can be carried out to look for human vulnerability. These tests try to convince employees to part with, for example, log-in details or sensitive data and information. This type of test evaluates the success or failure of the security policies, procedures and processes which have been implemented to protect against a social engineering attack. This type of test can uncover any weaknesses in employees' understanding of the security policies and procedures and may act as a catalyst for staff training.

The different types of social engineering are covered in Content area 4, section 4.1.3.

Physical penetration testing

This attempts to test the physical security in place. This type of penetration testing aims to test access to rooms or buildings (in an attempt to steal and/or remove digital devices, hard drives or recycling containers) to assess the effectiveness of the current physical security measures. As with social engineering, this type of test can reveal weaknesses in employees' understanding of the security policies and procedures and may trigger staff training.

Research

Two other types of security testing that can be carried out are web application and wireless. Research these types and create an infographic detailing how and why these are carried out.

White box testing

This is when the people carrying out the penetration tests are provided with full and complete information about the digital system to be tested. White box testing aims to identify any existing vulnerabilities in the software and any incorrect configurations within the digital system.

Black box testing

This is when the people carrying out the penetration tests are provided with no information except the name of the business or organisation. Black box testing is carried out from an external perspective with the aim of identifying ways that the digital systems could be accessed by attackers. The main disadvantage of using black box testing is that, because full and complete details have not been provided, vulnerabilities within the digital system may not be identified.

Test yourself

1 Who are the NCSC?
2 What is ethical hacking?
3 What does network penetration testing aim to test?

4 Identify two other types of penetration testing.
5 What is the difference between white box and black box testing?

Skills practice

An online games company provides games to its customers. Customers need to register their personal and payment details to buy and play the games.

Different options are available for the games. Some games are played online with the players' progress being stored on a dedicated games server. Other games are available to buy and download, meaning they can be played offline.

Some of the games are single player while others can be played by several players at once. If several players want to play the same game at once, each player must be a registered user.

Each customer has a username and password. The password is provided when registration is completed. The players can change their password to something more memorable. The username and password are stored on the games company's server.

The games company has been the victim of several data breaches and threats, including a DDoS attack, malware and social engineering attempts.

You have been asked to:
▶ Explain to the owner of the games company the importance of maintaining the CIA triad relating to customers' personal and payment details.
▶ Provide details about the threats that have happened and other potential threats to the games company.
▶ Explain the possible human threats to the digital systems, data and information stored by the games company.
▶ Provide details of possible processes and procedures that could be implemented to mitigate against future attacks, including recommendations.

Assessment practice

1 Explain why it is important to maintain the confidentiality of employees' salaries.
2 Discuss the financial impacts of an organisation failing to maintain privacy and confidentiality of its customer data and information.
3 Explain what is meant by a man-in-the-middle attack.
4 Identify and describe two different types of social engineering.

5 Explain the threats a malicious employee could pose to an organisation's digital system, data and information.
6 Explain the CIA triad (triangle).
7 Define hashing and asymmetric encryption, explaining the difference between them.
8 Discuss how user access restrictions can be used to mitigate against threats and ensure security.
9 Compare the use of a password and a passphrase.
10 Explain the process of location-based multi-factor authentication.

Types of assessment

This qualification is assessed in several ways.
► Core Component:
  • Core Papers 1 and 2
  • Employer-set Project
► Occupational Specialist Component:
  • synoptic assessment

About the Core Component assessment

There are three assessments for the core components of the Pearson T Level Technical Qualification in Digital: Digital Production, Design and Development.

There are two traditional exams (Papers 1 and 2) and an externally set Core Project. Each of the three components will contribute 33.33% towards your final Core Component grade.

Exams

The two exams are:
► Paper 1: Digital Analysis, Legislation and Emerging Issues
► Paper 2: The Business Environment

Each of the exam papers covers different content areas.

Paper 1	Paper 2
1 Problem solving	5 Business context
2 Introduction to programming	6 Data
3 Emerging issues and impact of digital	7 Digital environments
4 Legislation and regulatory requirements	8 Security

Each exam lasts for 2 hours 30 minutes and is worth a total of 100 marks.

This may sound like a long time, but time goes very quickly when you are doing an exam.

All exam questions use a command or keyword, such as 'identify' or 'describe'. You must recognise these because the words determine what you are required to do to be awarded the allocated marks. The allocated marks are shown in brackets ( ) usually at the end of the question.

The command words determine what is required, so if you understand their demands, it will help you to formulate your answer.

State, give, identify

You should answer these questions with a single word or phrase. These questions are low demand and are usually worth 1 mark per answer required. For example:

> Identify **two** different types of testing. (2)

Some of these types of questions will need a specific number of answers. In the example question, you will see that the two is in bold. This means you need to provide two answers. There may be numbers 1 and 2 on the answer lines to help you structure your answer.

Describe

This is moving to a higher level of demand. These answers are usually allocated 2 marks, but sometimes more. If a context is given in the question, you need to provide an answer that matches this context.

Identify and describe

Identify and justify

Identify and explain

These questions are asking you to do two steps in your answer. The first step is to identify, and the second step is to describe, justify or explain what you have just identified.

Justification means giving reasons for your identification.

You need to provide a correct identification before you can be considered for the marks allocated for the rest of the question. For example:

> Identify **one** layer in the TCP/IP model explaining its function. (4)

In the example question, the number **one** is in bold. This means you have to identify one TCP/IP layer. If this answer is correct, then your answer explaining the function of this layer can be considered for marks. In this example question, the explanation would be worth 3 marks.

Explain

This is moving to a higher level of difficulty than a describe question. These questions can be allocated 2 or 3 marks, but sometimes more. If a context is given in the question, you need to provide an answer that matches this context. For example:

> Explain **one** benefit and **one** drawback of connecting devices to form a network.   (6)

This question requires an answer that focuses on a benefit and a drawback of connecting devices to form a network. There are 6 marks available for this whole question. Therefore, it is logical to assume that 3 marks are available for the benefit and 3 marks for the drawback. You may find the words 'benefit' and 'drawback' on the answer lines. As with numbers, this will help you to structure your answer.

Compare

For this command/keyword you will need to write about the two different alternatives provided in the question.

The most common mistake on 'Compare' questions is to write about one of the alternatives in one paragraph and the second alternative in a different paragraph. To be considered for the marks available it must be clear that you have made comparisons. Use words such as: 'however', 'and', 'but'.

Discuss/evaluate/analyse

These command/keywords require a structured extended answer. They can be allocated 6 to 12 marks. Depending on the question, you may need to consider different viewpoints and ideas, as well as strengths and weaknesses or benefits and drawbacks.

The question may ask that two different aspects are specifically included in the answer; to maximise the marks for this it is important that you include both aspects.

An analysis requires an answer that breaks down an idea, usually provided in the question, into component parts. Each component part will need to be considered before a conclusion is reached.

These types of questions will be marked using a Levels of Response mark scheme. This means you will get marks for the depth of your answer and the application

of the knowledge and understanding to the context of the question.

Examples

In some questions you may be asked to provide an example in your answer. In this case there is a high probability that the example will have a mark(s) allocated to it. It is important that any example you provide in your answer must be appropriate to the context of the question.

If the question context was that of security of a college and an example was required, an example relating to health would not be appropriate and you would not gain the allocated mark(s) for the example.

There are **other types of questions** that may be included in either of the exam papers. Some of these types are:

Tables

You may be asked to complete a table in either of the exams. It is important that you answer in the appropriate area of the table. Examiners will only award marks if the answer is in the correct place. Each correct answer in a 'complete the table' question is generally worth 1 mark.

Diagrams

You may be asked to draw a diagram, for example, a flowchart, in the exams. It is probably best to stop and think before you start the draw the diagram. It is also important to draw the diagram asked for in the question. If you make a mistake, then either cross it out neatly or start a new diagram on the additional pages or extra paper available.

Code/pseudocode

It is possible that in Paper 1 you may be asked to create some code or pseudocode. As with the diagrams, space will be given in the exam paper for you to use when formulating your answer. It is a good idea to plan your answer to these types of questions before you start to write it down. If you do plan, then make it clear which is the plan and which is the actual answer that needs to be marked.

Exam hints and tips

▶ Always read the instructions carefully, that is what is the command/keyword asking for, for example identify, select, state, describe, explain, discuss, analyse, evaluate, justify and so on.

- Concentrate on one question at a time and ask yourself:
  - Do I understand what the question is about?
  - How many marks is the question worth? You should try to work to one minute per mark.
  - How many parts are there to the question?
  - Can I provide a well-constructed answer?
  - How am I going to answer the question?
  - Do I need to include examples?
  - Do I need to relate my answer to a particular context?
  - Do I need to use technical terminology?
- It is important that the person marking your paper can not only read your handwriting but can also understand what it is you are trying to tell them.

If they cannot read it or understand it, they cannot award you marks.

- If you make a mistake, cross it out neatly and then start again. There may be extra pages at the back of the exam paper, or you can ask for extra paper. If you use the extra pages or paper, then you must make it clear where your answer can be found.
- When you have finished answering the questions, and if you have time, go back over your answers. Read carefully what you have written and ask yourself:
  - Have I answered the question?
  - Have I answered all parts of the question?
  - Have I met the demands of the command/keyword, for example have I explained?
  - Have I used the correct technical terminology?

Core Component Project

The core component project is externally set and focuses on creating a solution for a business. This project is referred to as an **Employer Set Project – ESP**.

The project includes a familiarisation task and five assessed tasks. You may be provided with data files which can be used during the completion of the five assessed tasks. For example, you may be provided with a blank test table or relevant data.

Familiarisation task

The familiarisation task will be issued one week before you start work on the assessed tasks. The aim of the familiarisation task is to enable you to complete research relating to the industry sector the business is included in. You will be given a task brief which provides details about the business and the business sector. The brief will also provide details about different areas that should be considered during your research.

This task should take about 6 hours and will enable you to become familiar with the ways that digital tools and technologies are used in that industry sector.

During the time allocated for completion of this familiarisation task you will need to carry out research. You will be able to access the internet, work with others in a group and make notes about the results of your research. But, you will not be able to use your notes when you begin the assessed tasks. You may need to draw on the knowledge you developed while studying Content area 5: Business context.

When carrying out the research, it is important that you do not get distracted and just focus on the business sector that the ESP is applicable to. One strategy you could use to focus your research is to read the brief and then plan what needs to be covered in your research.

The brief will provide an overview of the business and the digital system that will be the focus of the assessed tasks. It is probable that the aims and requirements of the digital system will also be provided in the brief. There may also be some suggested considerations provided in the brief.

The aims, requirements and suggested considerations should be the starting point for you planning your research.

You are not required to submit for assessment any of your findings from your research.

Assessed tasks

There are five assessed tasks which will need to be completed. These are shown in this table.

Task number	Focus	Number of allocated hours	Marks allocated
1	Planning a project	3	19
2	Identifying and fixing defects in existing code	3	21
3	Designing a solution	3	17
4a	Developing a solution	4	34
4b	Reflective evaluation	1.5	9

Each task will have a brief. You will not be able to access any webpages during the controlled hours allocated for any of the tasks. You will be able to use offline versions of any software to produce your evidence.

Each task brief will include a section relating to the outcomes needed for submission. These are the outcomes that are needed so that your evidence can be assessed. You will be told which file format to use when saving your evidence, and the file name convention which must be used when you save your evidence to your assessment evidence folder.

It is vitally important that you follow these instructions and do not deviate from them at all. If you fail to follow the instructions on the submission of your evidence, your evidence may not be marked.

Task 1 Planning a project

Task 1 focuses on planning the project. You will be given a task brief which will provide information to help you complete the task and the outcomes that are required. You will be provided with further information that will help you to create the required planning outcomes. It is very important that you read all parts of the task before you start to plan. When planning, it is important that you consider all relevant information.

It is possible that you will be provided with a range of details to be considered during the planning of the task. These may include:
▶ financial information for the business, including a breakdown of projected costs for the development of the new system
▶ the project development team, including their names and job titles, skills and costs per hour
▶ a breakdown of the hours allocated for each task in the development
▶ a list of the assumptions that have been made when creating the development breakdown
▶ the total time allocated to the development.

There will be three main requirements for the planning including:
▶ producing a Gantt chart
▶ a plan relating to the required resources and the costs of these
▶ a rationale which explains the planning approach you took, including justifications for the decision you made.

There are 3 hours allocated for the completion of Task 1. So, a good starting point to ensure you do not run out of time could be to allocate 1 hour to each requirement. Each requirement of Task 1 has different marks allocated. You will also be able to demonstrate a range of core competencies. The marks and competencies for Task 1 are shown in this table.

Requirements	Marks allocated	Competencies covered
Gantt chart	6	E5 Synthesise information M1 Measure with precision D1 Use digital technology and media effectively
Resource and cost plan	4	M8 Communicate using mathematics
Rationale	9	E5 Synthesise information M9 Cost a project

Gantt chart

When you are creating your Gantt chart you will need to:

- ▶ consider the strengths and skills of the development team and assign appropriate tasks to them – the details of the development team may have been provided in the task brief
- ▶ make scheduling decisions in response to a defined deadline – the projected breakdown of hours and the development time will have been provided in the task brief
- ▶ prioritise activities or tasks based on analysis of requirements – you will need to consider **milestones**, **contingency time**, **dependent**, **serial** and **concurrent tasks**, and any task dependencies
- ▶ assign resources to project tasks
- ▶ organise the tasks efficiently.

Resource and cost plan

When you have completed your Gantt chart, you will need to create a plan which covers the resources you will need and their costs. This plan should be created in an Excel compatible spreadsheet. The plan should include:

- ▶ required and allocated resources, including which team member(s) will be involved in each task
- ▶ the cost of the resources
- ▶ an overall estimate of the total cost of the development.

> ### Key terms
>
> **Milestones:** specific points during a project used to make sure it is on-track. A milestone has no time allocated and usually occurs at the end of a critical part of the project. It is usually shown as a diamond shape on a Gantt chart.
>
> **Contingency time:** time built into the schedule for a project which can be used in case of any unforeseen issues, problems or events.
>
> **Dependent tasks:** tasks which cannot be started until a previous, specified task has been completed.
>
> **Serial/concurrent tasks:** tasks which can be completed at the same time.

When you are creating this plan, you should consider the skills and industry experience of the development team and how much experience they have. You will

need to use the costs for each development team member that were provided in the task brief.

In the information in the brief, different options may have been provided, for example different server options which can be considered. When completing the cost, you will need to consider which option you will select from those provided. You should also consider any ongoing costs related to the final digital system. For example, the cost of any staff needed or maintenance costs.

Rationale

When you are selecting any options, you will need to be able to justify your choice as this justification will form part of the final part of this task – the rationale.

The rationale will explain the planning approach you took including justifications for any decisions you made. When developing your rationale for the decisions you made during the planning you need to consider:

- ▶ costs, including the cost of the development and any ongoing costs, the risks and benefits including how you have mitigated against the risks
- ▶ the order and timing of tasks including any milestones, serial/concurrent and dependent tasks and contingency time
- ▶ selection and allocation of resources, for example development team members and any physical resources required.

Task 2 Identifying and fixing defects in existing code

The focus of Task 2 will be to identify and fix defects in existing code. As with Task 1, you will be given a task brief which contains the task's requirements and the required outcomes. For this task you will be given some code which does not function as intended and a test log that must be used.

Before you start identifying the errors in the non-functioning code, it is, again, very important that you read all parts of the task. This is to help you understand what is required from you and also from the non-functioning code you will amend.

You may need to draw on the knowledge you developed while studying Content area 2: Introduction to programming.

You will need to use the provided test plan and the non-working code files when you generate evidence for this task.

There will be two main requirements for this task which are:

► using testing to identify defects with the non-working code and documenting the testing process including any possible remedial actions to be taken

► producing a working solution based on the results of your testing.

There are 3 hours allocated for the completion of Task 2. So, a good starting point to ensure you do not run out of time could be to allocate each requirement 1.5 hours. Each requirement of Task 2 has different marks allocated. You will also be able to demonstrate a range of core competencies. The marks and competencies for Task 2 are shown in this table.

Requirements	Marks allocated	Competencies covered
Use of testing to identify defects, and documenting the testing process	12	E1 Convey technical information to different audiences
		E2 Present information and ideas
		M10 Optimise work processes
		D3 Communicate and collaborate
A working solution	9	M4 Use rules and formulae
		M5 Process data
		M7 Interpret and represent with mathematical diagrams

The brief will provide you with the requirements of the code. The requirements could be user requirements, for example, user data-entry requirements, and/or requirements set by the business, for example, different financial options.

Use of testing to identify defects, and documenting the testing process

When you have read and understood the defined requirements you will need to analyse the non-working code and create a test plan to identify the issues that are causing the code to be non-working. You will need to draw on the knowledge from Content Area 2: Introduction to programming, section 2.8: Testing.

When you are using testing to identify defects in the non-working code you will need to:

► assess the given code against requirements
► carry out testing to identify issues in the given code
► perform any remedial actions required, justifying any decision made when fixing the defect.

The test plan you create, use and amend should cover:

► Identify the tests to be carried out.
► Describe the purpose of the identified test.
► Identify the test data to be used including valid, valid extreme, invalid, invalid extreme and erroneous.
► Describe the expected results.

► Describe the actual results of the tests performed.
► Compare the actual results of testing with the expected results.
► Describe any further actions that are required.

At this stage it is important that you also test the code to make sure it is robust and performs as required. One way of doing this is to do a dry run with values that should produce an error message and that this error message only appears when incorrect values have been input.

The solution

You will then need to make the changes to the code and confirm the changes you have made make the code work as required.

You should look for formatting errors, for example indentation errors, as well as functionality errors in the non-working code. For example:

► if the code requires a date output, checks should be made that the defined format is that of the UK date format – dd/mm/yyyy
► any currency outputs are in UK £ with 2 decimal places.

You must follow accepted programming conventions so you will need to draw on the knowledge from Content Area 2: Introduction to programming.

When you are making the changes to the code you must keep referring back to the requirements in the brief so you can ensure you are meeting these. You should consider:

▶ the correction of errors in the code including the addition and/or deletion of code to ensure that the code is functional and meets the specified requirements

▶ following appropriate programming conventions when fixing code to ensure that it makes use of precise logic and programming structures throughout, so that the program produces consistently correct outcomes.

Task 3 Designing a solution

Task 3 will focus on designing a solution to a given set of requirements and the specified required outcomes. These outcomes, as with all the other tasks, will form the basis for the evidence you are required to produce. You will be provided with data files which you must use when you are designing your solution.

In this task you have to produce algorithm(s) which meet the defined user requirements. You will be told which types of tools are acceptable to produce your algorithm(s); these are likely to be pseudocode/flowcharts. However, you should also aim to create a visualisation of the decomposition of the problem.

The knowledge you developed while studying Content area 1: Problem solving will help you to carry out this task.

There will be one main requirement for this task, which is:

▶ algorithm design(s).

There are 3 hours allocated for the completion of Task 3. The requirement for Task 3 has different marks allocated for different aspects. You will also be able to demonstrate a range of core competencies. The marks and competencies for Task 3 are shown in this table.

Requirements	Marks allocated	Competencies covered
Decomposition of the problem	8	D4 Process and analyse numerical data
Application of logical thinking and conventions	6	M2 Estimate, calculate and spot errors
		D2 Design, create and edit documents and digital media
		E3 Create texts for different purposes and audiences
Communication of the design	3	E4 Summarise information/ideas
		D3 Communicate and collaborate

Decomposition of the problem

When you are decomposing the problem, you should:

▶ break down the problem into smaller parts, justifying any decisions made, and use decomposition to show all the necessary subsystems that make up the main solution

▶ use elements of reusable components

▶ visualise the decomposition, using appropriate tools to communicate algorithms (e.g. flowcharts, pseudocode, data flow diagrams).

Application of logical thinking

You also need to demonstrate that you can apply logical thinking by:

▶ describing the parts of the solution using algorithms, justifying how these algorithms form a complete solution to the problem

▶ clearly and uniquely defining the steps – each step should depend on the input and the result of the preceding steps

▶ ensuring the algorithm makes use of key constructs (e.g. sequence, selection and iteration).

Use of conventions and communication of the design

It is important that you consider and use relevant and applicable conventions, and make sure that the design documents contain a high level of detail.

This will include:

▶ The correct use of structure and convention based on your selected algorithmic designs, for example flowcharts, pseudocode, data flow diagrams, module diagrams. For example, the correct and consistent use of symbols for flowcharts and keywords used in pseudocode. This can also include the use of indentation, structure, notation and syntax.

▶ The selection and consistent use of appropriate naming conventions throughout. This includes using sensible names and keywords. The names you choose should enable a third party to understand your algorithmic designs.

▶ The effective communication of your proposed solution which will allow the client to make informed decisions and enable a third party to use design documents to create the proposed solution, which will include the appropriate combination of written and diagrammatical presentation.

When you are creating your algorithmic designs, you should also consider:
▶ appropriate use of technical vocabulary
▶ the audience
▶ explanations of structures and process in the design.

Task 4a Developing a solution

Task 4a will require you to develop a solution. The requirements of the, usually, additional functionality to the solution you designed and manipulated in Task 3, will be given to you. The task brief will typically provide extra system requirements and user requirements.

It is very important that before you begin to code your solution, you read and understand both the system and user requirements. By considering these requirements

before you start to code, there is a greater chance that your solution will meet these requirements.

While you are coding it is also worth checking back to these requirements regularly to make sure you have not deviated from them. You will usually be provided with a data file which can be used to test your solution. As with the other tasks you will be told the file naming convention that you must use and which file format(s) to use when uploading your evidence.

You may need to draw on the knowledge you developed while studying Content area 2: Introduction to programming.

There will be one main requirement for this task. This is:
▶ developing code to meet the defined requirements.

There are 4 hours allocated for the completion of Task 4a. The requirement for Task 4a has different marks allocated for different aspects. You will also be able to demonstrate a range of core competencies. The marks and competencies for Task 4a are shown in this table.

Requirements	Marks allocated	Competencies covered
Functionality	6	
Logic and programming structures	3	
Robustness	3	
Security	6	
Code organisation	8	M10 Optimise work processes
User experience	8	E6 Take part in/leading discussions

When you are creating your code, it is important that you keep referring back to the specified requirements provided in the brief. It is probable that the requirements will include system and user requirements. If you deviate from either of these sets of specified requirements, then it is unlikely that your solution will be acceptable to the end user.

When you are creating your coded solution, you must:
▶ Consider the security requirements and use secure coding principles and practices to mitigate against potential threats and vulnerabilities
▶ Ensure your code is maintainable, readable and functional, and follows accepted programming conventions.

You will not be required to provide formal evidence that you have tested your code. However, to ensure that the code meets the specified system and user requirements you must test the code as you create it.

Your final code must:
▶ be well-structured including the use of modules with a defined structure, for example separated modules, the use of procedures, functions or classes
▶ have clear annotation to enable future maintenance
▶ use sensible and appropriate naming of all variables and structures.

Your code should also meet the defined user requirements. This could be demonstrated by the inclusion of:
▶ user input handling including the use of validation
▶ user guidance and error messages
▶ outputs that meet the specified requirements, the end user's needs and that are fit for purpose.

Task 4b Reflective evaluation

Task 4b focuses on a reflective evaluation of the code you created in Task 4a. For this task you will be given access to read-only copies of the code you created in

Task 4a. This will allow you to consider the code when creating your evaluation. The task brief you will be given will include the systems and user requirements from Task 4a.

There will be one main requirement for this task which is:

▶ a reflective evaluation.

There are 1.5 hours allocated for the completion of Task 4b. The requirement for Task 4b has different marks allocated for different aspects. You will also be able to demonstrate a core competency. The marks and competencies for Task 4b are shown in this table.

Requirements	Marks allocated	Competencies covered
Programming outcomes	6	
Comparison to designs	3	E3 Create texts for different purposes and audiences

During your evaluation you will need to consider how your solution met the system and user requirements and how the solution could be developed further. You should make sure that you use, and provide, specific examples for the brief and your code to exemplify the points you are making.

An effective reflective evaluation should consider:
▶ how and why the solution meets the specified system and user requirements
▶ how and why the solution could be further developed.

You should make sure you include specific examples to support the points you make in your evaluation.

A last note

Examiners who will mark your exam papers, and moderators who will assess the evidence you provide for your ESP, are essentially nice people who would like to give you marks. But they cannot read your mind. So, it is really important that you make sure that everything you write, in your exam and the ESP, is clear and unambiguous.

Good luck with your exams and the ESP.

References

BBC (2022) *Data, information and knowledge.* Available at: https://www.bbc.co.uk/bitesize/guides/zkfbkqt/revision/4 Accessed: May 2022

BCS (2022a) *About us.* Available at: https://www.bcs.org/about-us/ Accessed: May 2022

BCS (2022b) *Code of Conduct.* Available at: https://www.bcs.org/membership-and-registrations/become-a-member/bcs-code-of-conduct/ Accessed: May 2022

Cybsafe (2020) *Human error to blame for 9 in 10 UK cyber data breaches in 2019.* Available at: https://www.cybsafe.com/press-releases/human-error-to-blame-for-9-in-10-uk-cyber-data-breaches-in-2019/ Accessed: May 2022

Department for Digital, Culture, Media and Sport (2018) *Data Protection Act 2018 Factsheet – Overview.* Available at: https://assets.publishing.service.gov.uk/government/uploads/system/uploads/attachment_data/file/711162/2018-05-23_Factsheet_1_-_Act_overview.pdf Accessed: May 2022

Equality and Human Rights Commission (2020) *Your rights under the Equality Act 2010.* Available at: https://www.equalityhumanrights.com/en/advice-and-guidance/your-rights-under-equality-act-2010 Accessed: May 2022

ICO (2019) *Introduction* to the Data Protection Bill. Available at: https://ico.org.uk/media/2614158/ico-introduction-to-the-data-protection-bill.pdf Accessed: May 2022

IETF (2004) *About: Mission and principles.* Available at: https://www.ietf.org/about/mission/ Accessed: July 2022

ISO (2022) *About us.* Available at: https://www.iso.org/about-us.html Accessed: May 2022

Oxford University Press (2022) "internet, n." OED Online. Available at: https://www.oed.com/ Accessed: May 2022

ScienceLogic (2022) *Data Modeling.* Available at: https://sciencelogic.com/glossary/data-modeling Accessed: July 2022

Softinterface (2022) *Fixed Width Text File Definition.* Available at: https://www.softinterface.com/Convert-XLS/Features/Fixed-Width-Text-File-Definition.htm

W3C (2022) *Leading the Web to Its Full Potential.* Available at: https://www.w3.org/ Accessed: May 2022

W3C (2004) *Architecture of the World Wide Web, Volume One.* Available at: https://www.w3.org/TR/webarch/ Accessed: May 2022

Glossary

Abstraction The process of removing or filtering characteristics that are not needed, in order to focus on essential characteristics.

Access rights Control over which user has access to a digital system, folders, files and data/information.

Accounts payable Money that is paid out by the organisation.

Accounts receivable Money that is coming into the organisation.

Active cooling Using fans to reduce the heat of computer components.

Active matrix Controls each pixel using capacitors. This enables the pixels to change colour and brightness more rapidly.

Active sensors Sensors requiring an external signal or a power signal.

Agency worker A person who has a contract with an agency but works temporarily for someone who hires them.

Analogue sensors Produce a continuous output signal relating to the quantity being measured.

Application Specific Integrated Circuit (ASIC) A microchip designed for special applications such as a handheld computer.

Automated teller machine (ATM) A special computer that allows a bank account holder to manage their account.

Backbone networks The part of a network that combines different networks into a single complete network. The backbone carries the bulk of the network traffic.

Big data Very large data sets that can be analysed to produce information such as trends and patterns. Big data cannot be analysed using traditional analysis data tools.

Black box testing Testing the software when the internal structure and design is not known to the test.

Bottleneck Limitation of data flow by network resources.

Bus topology All nodes are connected directly to a central cable that runs up and down the network – this cable is known as the backbone. Data is sent up and down the backbone until it reaches the correct node.

Capital The money or assets owned by a person or organisation and used as investments to make more money.

Cathode Ray Tube (CRT) A vacuum tube containing an electron gun at one end and a fluorescent screen at the other end.

Chargeback A payment amount that is returned to the debit or credit card of a customer after they successfully dispute a transaction or return a purchased item. The chargeback can be initiated by the seller or the customer's bank. Sellers are often charged a fee from the card issuer when a chargeback occurs.

Confidentiality, integrity and availability (CIA) Also known as the CIA triad.

Code of conduct A document which defines rules, values, ethical principles and vision.

Cognitive development How we think, explore and work things out. It is the development of our knowledge, skills, understanding and ability to solve problems.

Competition law The purpose of this law is to promote healthy competition. It makes it illegal for anticompetitive agreements to be in place between two or more organisations, for example to share markets and fix prices. It also makes it illegal for businesses to abuse their dominant market position.

Computational thinking A problem-solving method using computer science techniques, where possible solutions are developed and presented in a way that can be understood by humans and computers.

Constants Values in a program that does not change when the program is being executed or run.

Contiguous data Data that is stored in a collection of adjacent locations.

Contingency time Time built into the schedule for a project which can be used in case of any unforeseen issues, problems or events.

Control keys Provide cursor and screen control. Include four directional arrows, Home, End, Insert, Delete, Page up, Page down, Control (Ctrl), Alternate (Alt) and Escape (Esc).

Creative Commons (CC) International, not-for-profit organisation that provides free licences for creators to use when making their work publicly available. The licences provide permission for others to use the work under certain conditions.

Cross-site scripting (XSS) Usually found in websites and/or web applications that accept end-user input. This can include search engines, login forms, comment boxes and message boards.

Cyber security The practice of defending computers, servers, mobile devices, electronic systems, networks and data from malicious attacks.

Dark Net Networks that are not indexed by search engines. They are only available to a select group of people with authorisation, specific software and configurations.

Dashboard A type of graphical user interface providing simple visualisation of data related to performance indicators. It is commonly accessible by a web browser and can show real-time data updates.

Data Raw facts and figures before they have been processed.

Data at rest Data stored on a digital device or storage medium.

Data cleaning The process of going through data looking for errors and correcting them, or excluding data where errors have been located.

Data encryption software Software that is used to encrypt a file or data.

Data in transit Data being sent to one or more authorised users.

Data Over Cable Service Interface Specification (DOCSIS) A globally recognised telecommunications standard. It supports high-bandwidth data transfer via existing coaxial cable systems that were originally used for the transmission of cable television programme signals (CATVS).

Data packets Small units of data which are sent and received when accessing the internet or any other type of network.

Data silos A group of raw data accessible by one department but not available to the other departments within the organisation.

Data sprawl The vast amounts and variety of data produced by organisations on a daily basis.

Data striping The technique used to store consecutive segments of data (e.g. a file) on different physical storage devices.

Data subject The person the data is being held about.

Data transfer protocol (DRP) The technique used to store consecutive segments of data (e.g. a file) on different physical storage devices.

Data virtualisation Connects all types of data sources regardless of the file types and locations. The data is then combined, and users can access the combined data through reports, mobile apps, websites, dashboards and portals.

Data visualisation The graphical representation of data.

DBMS database management system.

Decomposition Breaking a complex problem into smaller sub-problems.

Dependent tasks Tasks which cannot be started until a previous, specified task has been completed.

Developmental The development of someone, something or even both.

Dictionary An iterable data structure that is built into Python. A Python dictionary has a series of 'keys' that have 'values' or 'data', so the dictionary is a set of mappings from specific keys to specific values. Unlike a dictionary we use for checking our spelling, a Python dictionary is unordered. In other words, the items in the dictionary are not in a set order. You can create an empty dictionary or a dictionary containing keys and values. Keys can be added, deleted and amended within the dictionary.

Digital sensors Work with discrete digital data. The digital data is used for conversion and transmission.

Distributed denial of service (DDoS) When a network is flooded with so much traffic that it cannot operate or communicate as required.

DPP Director of Public Prosecutions.

DSE Display Screen Equipment.

Encapsulation Information is taken from a higher level and a header is added to it, treating the higher layer information as data. The IP packet is then encapsulated into a Layer 2 Ethernet frame. The frame is then converted into bits at Layer 1 and sent across the local network.

Encryption code/key A set of characters, a phrase or numbers that are used when encrypting or decrypting data or a file.

End user A consumer of a product and/or service. This does not only apply to customers/clients but also to employees.

Ethical hacking An alternative term for penetration testing.

Ethics The moral principles, or rules, that govern a person's attitudes and behaviour.

External bus Also known as an external bus interface (EBI) or expansion bus. It is a type of data bus that enables external devices and components to connect with a computer.

External stakeholders Groups outside an organisation, for example shareholders.

Facial recognition software Software that can identify or confirm someone's identity using their face in a photo, video or in real time.

Field Programmable Gate Array (FPGA) An integrated hardware circuit that can be programmed to carry out one or more logical operations. Uses include the rear viewing cameras on cars, data analytics, encryption, compression and Artificial Intelligence (e.g. within deep neural networks).

Filter noise Noise is unwanted electrical or electromagnetic energy that degrades the quality of signals and data. It occurs in digital (and analogue) systems. It can affect files and communications including, text, images, audio and so on. The filter part is finding it and getting rid of it (or at least reducing it).

Firmware Code, added at the time of manufacturing, written to a hardware device's non-volatile memory. It is the software that allows the hardware to run.

Foreign key This is used to link tables together. A foreign key is a field in one table that is linked to a primary key in a different table.

Frequency response The measurement rate of the highs and lows of the sounds produced by a speaker.

Function keys Arranged in a row at the top of the keyboard, they are assigned a unique meaning and used for a specific purpose.

Hacker A person who uses computers to gain unauthorised access to data.

Haptics Using technology to stimulate the senses of touch and motion to reproduce the sensations that would be felt by someone interacting directly with the physical object.

Hash A number generated from a string of text.

Hops Refers to the number of routers that a packet passes through from its source to its destination. A hop can also be counted when a packet passes through other hardware on a network such as switches, access points and repeaters. It is dependent on what role the devices have on the network and their configuration.

HTML Hyper Text Markup Language.

HTTP Hypertext Transfer Protocol.

Humidity The amount of water vapour in the air. The higher the humidity, the more water vapour there is in the air.

Hypervisor Software that creates and runs virtual machines (VMs). A hypervisor, which is sometimes referred to as a virtual machine monitor (VMM), isolates the hypervisor operating system and resources from the virtual machines. This allows it to create and manage the VMs.

ICO Information Commissioner's Office.

Immutable When the value of a number data type changes, a new object is created.

Index A numerical representation of an item's position in a sequence.

Information Data + [structure] + [context] + meaning

Injection flaws Allow attackers to relay malicious code through an application to another system.

Integrated Development Environment (IDE) An application containing tools and functions used by developers to create software, and which can often support multiple languages. It has a runtime environment enabling developers to execute a program line by line which is useful for testing code

Internal stakeholders Groups within an organisation, for example owners and employees.

Internet Protocol (IP) The string of numbers an ISP assigns a device.

IP spoofing Changing a packet's source IP address to impersonate another computer system, or to hide the identity of the sender, or both.

Inventory Consists of finished products and assets owned by an organisation or used by an organisation to carry out production.

Iterative/iteration The repeating of a task.

Jitter When there is a time delay in the sending of data packets over a network.

Knowledge The ability to use information to, for example, form judgements and make decisions.

Limited Liability Partnership (LLP) A company owned by two or more people. Each person pays tax on their share of the profits but is not personally liable for any debts the company cannot pay.

Lumens The amount of light output (lms). These are a measure of the total amount of visible light (to the human eye) from a lamp or light source. The higher the number of lumens, the brighter the lamp will appear.

Maintenance, repair and operations (MRO) supplies The maintenance, repair and operations of equipment and machinery.

Malware Malicious software.

Media Access Control (MAC) This is a code which is in the device's Network Interface Card (NIC), identifying the physical device.

Metrics A set of numbers that gives information about a particular process or activity.

Middleware Software which is 'in the middle' of the operating system and the applications working on it. It allows communication and data management for distributed applications by operating as a hidden translation. It is used to link two separate applications together.

Milestones Specific points during a project used to make sure it is on-track. A milestone has no time allocated and usually occurs at the end of a critical part of the project. It is usually shown as a diamond shape on a Gantt chart.

Mineral oil cooling Computer components are submerged in mineral oil. Heat generated by the components is transferred to the mineral oil at a better rate than air. The mineral oil dissipates the heat.

Mitigate If you mitigate against something, you take steps to reduce the likelihood of it happening, or to reduce its impact if it does happen.

Morals The principles of what people believe is right or wrong.

NCSC The UK's National Cyber Security Centre.

Node A connection point in a network that can receive, send, create or store data. Each node requires some form of identification to receive access, such as an IP address. Examples of nodes are computers, printers, routers and switches.

Non-linear thinking The ability to make connections and draw conclusions from unrelated concepts or ideas.

Open source Software where the copyright holder grants users the rights to use, edit and distribute the source code to anyone and for any purpose.

Operational intelligence Data analysis that enables decisions and actions to be made in business operations based on real-time data as it is generated and collected. The data analysis process is automated and the results are integrated into operational systems for immediate use by the managerial staff and other relevant employees.

Opt in Means a person has to take a specific positive step, for example tick a box, send an email or click a button or icon, to say they consent to receiving marketing.

Opt out Means a person must take a positive step to refuse or unsubscribe from marketing.

Outage A period when the power, a service or equipment is closed down.

Packet A small segment of a larger message. Data that is sent over computer networks is divided into packets. The packets are combined by the computer/device that is receiving the message.

Packet loss When one or more packets fail to reach the intended destination.

Pain point Issues that people will work around. In some instances the user is not even aware they are happening.

Parameter A special kind of variable in computer programming languages that is used to pass information between functions or procedures.

Parity A technique that checks whether any data has been lost or overwritten when it is moved from one storage place to another or transmitted between computers on a system.

Parse The formal analysis by a computer of a sentence/string of words into its constituent parts.

Passive cooling The cooling of computer components by slowing the speed at which the component is operating.

Passive matrix Uses a grid of vertical and horizontal wires to display an image on the screen. Each pixel is controlled by an intersection of two wires in the grid. When the electrical charge is altered at a given intersection, the colour and brightness of the corresponding pixel can be changed. Passive matrix is relatively simple and inexpensive to produce, but the disadvantage is that the charge of two wires (vertical and horizontal) must be changed in order to change just one pixel. The response time is therefore slow. Fast movement may appear blurry or faded.

Passive sensors Do not require external power signals and directly generate an output response.

Peer/peering The arrangement of traffic between Internet Service Providers (ISPs). The large ISPs which have their own backbone networks allow traffic from other large ISPs in exchange for traffic to be allowed on their backbones. In addition, they exchange traffic with the smaller ISPs so that regional end points can be reached.

Performance metrics The process for the collection, analysis and reporting of information with respect to the performance of a component, system, organisation, department or individual.

Permissions A list of attributes that determine what a user can do with files and folders, for example read, write, edit or delete.

Persona Fictional characters, based on research, created to represent the different types of end user that may access the products/services in a similar way.

Personal data Any information relating to an identified or identifiable living individual

Platform agnostic Where software is able to work on all devices and can operate across a variety of operating platforms.

Primary key A field in a table that allows each record to be uniquely identified. For example, every person 16 years or older in the UK has a National Insurance number. This uniquely identifies a person.

Public domain Belonging to or being available to the public as a whole.

Qualitative data Non-numerical data.

Quantitative data Numerical data.

Radio frequency identification (RFID) Tiny chips that contain information which is transmitted when near a receiver.

Random access memory (RAM) Short-term data storage.

Rate latency The amount of time (delay) it takes to send information from one point to the next.

Read performance The time taken to open a file from storage.

Redundant/redundancy Where a system is still able to function regardless of issues that may occur.

RFC Request for Comments.

Runtime The length of time a computer program takes to run.

Search Examine data to find a specified value.

Segmentation Dividing a computer network into smaller parts.

Semantics The process followed when executing a program in a specific language.

Serial/concurrent tasks Tasks which can be completed at the same time.

Social listening The process of monitoring social media channels for mentions of the organisation's brand, competitors, products and so on. It gives organisations an opportunity to track, analyse and respond to conversations about them on social media.

Societal norms Unwritten rules about beliefs, attitudes and behaviours that are considered acceptable in a specific social group or culture.

Socioeconomic status A measure of a person's combined economic and social status (tends to be positively associated with health).

Sort Put a data set into a specified order.

Special purpose keys Include Enter, Shift, Caps Lock, Num Lock, Spacebar, Tab and Print Screen (PrtScn).

Stakeholder Anyone with an interest in a business or organisation. Stakeholders can be individuals, groups or other organisations, or businesses that are affected by the organisation's activity. They can include customers, suppliers, employees, communities, government and even the ecosystem.

Stock Consists of finished products, parts and materials which are sold to customers.

Subroutine Sometimes referred to as a routine, a function, a procedure or a subprogram, it is code that is called and executed anywhere in a program. An example is to display a message: instead of writing out the code each time, routines are created and called when required.

Syntax (in linguistic statements) The structure of statements.

Syntax (in programming code) A general set of rules for how words and sentences should be structured. These rules are known as the language syntax. When writing programming code, the syntax defines how declarations, functions, commands and other types of statements are arranged.

Thin Film Transistor (TFT) Used in high-quality flat display LCDs. There is a transistor for each pixel on the screen allowing the electrical current that illuminates the display to turn on and off at a faster rate. This makes the display brighter and motion smoother.

Time to live (TTL) The amount of time or 'hops' that a packet is set to exist inside a network before being discarded by the router. TTL is also used in Content Delivery Network (CDN) caching and Domain Name System (DNS) caching.

Total harmonic distortion (THD) The amount of distortion created by the signal amplification.

Trace table A tool used to test or dry run algorithms to make sure no logical errors occur while calculations are being processed. Each column represents a variable and the rows represent the numerical input and the output of the variable.

Transformational Producing a change or improvement in a situation.

Transitional The transition (movement) from one position, stage, state or concept to another.

URI Uniform Resource Identifier.

URL Uniform Resource Locator.

Validation Checks that the data being entered into a digital system is sensible and reasonable, and checks it against pre-set rules.

Variable (general) Values that will change usually as a result of an input or of a calculation being carried out.

Variable (in computing) A memory location in a program where values are stored.

Verification A check to see whether the data being entered into a digital system is identical to the source document or initial data entry.

Video random access memory (VRAM) A known as a dual-ported memory. It is an expensive form of RAM which has the capability to perform reads and writes simultaneously. VRAM can also be accessed by two devices at the same time and is commonly used to increase the speed of video cards.

Watts The total amount of amplification available for speakers.

White box testing A method of testing when the internal structure and design of the software is known to the tester (usually carried out by the software development team).

Wisdom The ability to use knowledge to perform an action.

Write performance The time taken to save a file to storage.

Acknowledgements

The publishers would like to thank the following for permission to reproduce photographs: page 1 © zinkevych/stock.adobe.com; page 6 © chrisdorney/stock.adobe.com; page 14 © Sergey/stock. adobe.com; page 55 © Tran/stock.adobe.com; page 58 © leungchopan /stock.adobe.com; age 66 © Andrey Popov/ stock.adobe.com; page 69 © elenabsl/stock.adobe.com; page 94 © BillionPhotos.com/stock.adobe.com; page 94 © vegefox.com/stock.adobe.com; page 100 © Gorodenkoff/ stock.adobe.com; page 101 © BethWolff43/Thinkstock/ iStock/Getty Images; page 102 © Xieyuliang/Shutterstock. com; page 105 © dreamnikon/Shutterstock.com; page 115 © vvoe/stock.adobe.com; page 124 © Aleksei Gorodenkov/ Alamy Stock Photo; page 131 © Natee Meepian / Alamy Stock Photo; page 135 © NicoElNino/stock.adobe.com; page 152 © Elegant Solution/stock.adobe.com; page 164 © Климов Максим/stock.adobe.com; page 208 © Alex/stock. adobe.com; page 226 © Andrea Danti/stock.adobe.com

Index